Big Game of Botswana

Clive Spinage

Big Game of Botswana

The Tragic History of a Once Great Southern African Fauna

Springer

Clive Spinage
Faringdon, UK

ISBN 978-3-031-60032-6 ISBN 978-3-031-60033-3 (eBook)
https://doi.org/10.1007/978-3-031-60033-3

This Springer imprint is published by the registered company Springer Nature Switzerland AG
The registered company address is: Gewerbestrasse 11, 6330 Cham, Switzerland

Preface

Consequent upon the publicity attending the mass mortality of wildebeest which took place in Botswana's Kalahari Desert in 1982, international attention became focused upon the country's environment and wildlife. Beset by one ecological problem after another, Botswana nevertheless saw an increasing growth in tourism, which grew even faster following the changed circumstances in South Africa. Promotional literature enthusiastically describes Botswana as one of Africa's last wild places, making use of such superlatives as "a wildlife enthusiast's paradise". But what we see today are but small remnants of scenes which once surpassed any other known faunal concentrations in Africa.

Botswana is fortunate in having many contemporary accounts of the country dating from the middle of the 19th century; fifty years before missionaries, explorers and sportsmen were to write in equal detail about East Africa. These early accounts show that Botswana formerly possessed a much greater and more widespread fauna, and the country was better watered than today. Vast areas of wilderness carrying only a sparse fauna today were once populated in abundance by big game.

By quoting contemporary observers, rather than being an exercise in nostalgia, this book aims to depict something of the former ecology of Botswana to provide a yardstick for comparison in a rapidly changing environment. When change is taking place so rapidly it is important to be able to understand what we are losing and what habitats *should* be like in the absence of man's recent destructive impact. It underlines the reality that the fauna of Botswana today is but a tiny remnant of the once great southern African fauna which blackened the plains to the Cape. It shows how the decline came about, whereas the pessimistic predictions at the end of the nineteenth century of extinction have happily not come true, the continued existence of this once great fauna is increasingly threatened. Its tragic history is seen through an anthology of destruction.

Faringdon, UK

Clive Spinage

Acknowledgment I am grateful to Brian Jackman for the permission to quote from Jackman, B. and Godwin, P. 1986. Wildlife driven to the wall in the Kalahari. *Wildlife*: 72–79.

Contents

List of Figures

List of Plates

Chapter 1
Background to Big Game

Abstract Quoting contemporary accounts Botswana described as the European hunters, traders, and missionaries saw it at the beginning of the nineteenth century; its animals and countryside. Tribes and their chiefs and government; drying up of the land. Chobe River. The Kalahari desert described. Abundance of game reported. Characteristic species described.

Keywords Big game · Botswana · History · Quotations

1.1 As It Was

Exhibitionist, thoughtless hunter, but of daunting courage, the Eton-educated laird described by Livingstone as a "mad sort of Scotchman", Roualeyn Gordon Cumming of Altyre, wrote of the Rhenoster (rhinoceros) river in 1848: "I found the boundless undulating plains thickly covered with game, thousands upon thousands chequering the landscape far as the eye could see in every direction. The blesboks, which I was most desirous to obtain, were extremely wary, and kept pouring on, on, up the wind in long-continued streams of thousands."

The South African trader, James Chapman, who was to hunt as far north as Botswana's present-day Chobe National Park, wrote a year later near the same area: "thousands of quaggas, blesbucks, springbucks and wildebeest are continually crossing the road and giving way at the approach of the wagons. We are all firing at these animals and notwithstanding the great numbers, we shot few indeed."

This was South Africa. A country still swarming with game twelve years after Captain Cornwallis Harris, on leave from India, hunted its grassy plains as far west as Parr's Halt on the east bank of the Limpopo River over the Botswana border, and announced to the sporting world the staggering profusion of animal life to be found. His enthusiastic books encouraged the headstrong Cumming to hunt, on horseback in a tartan kilt, in the country known today as Botswana. Soon followed by sportsmen like William Cotton Oswell, who was to lead Livingstone to the fabled Lake Ngami.

C. Spinage, *Big Game of Botswana*, https://doi.org/10.1007/978-3-031-60033-3_1

Cumming was not the first European that we know of to hunt in Botswana. By the 1820s some Afrikaner hunters had reached Lephephe, and Conrad Buys, a renegade from Cape Colony, moved to Shoshong in 1820, dying at Palapye some years later. Beginning his hunting along the Limpopo in 1809, he allegedly shot some 600 elephants. But an old servant of Khama I (c1795–c1817) informed the Cape government explorer, Dr. Andrew Smith, that the first white man to visit them was one who passed by and went down the Marico 30 to 40 years before. Perhaps the missionary the Reverend W. Edwards, who reached Ngwaketse country in 1807 or 1808.

In 1826 Andrew Bain hunted giraffes and white rhinoceroses in the south. Bain, accompanied by a Mr. Biddulph, was an early visitor to Molepolole, recording that he passed, after crossing the Molopo, through a country abounding with game, especially white rhinoceroses and giraffes. The Scottish missionary Robert Moffat, had reached Kanye in 1824, travelling with a party of Griquas, who went to hunt elephants in the area, but he tells us little of the land and its wild life from this visit. David Hume, yet another Scotsman, was the earliest recorded trader. In 1830 he followed the Mahalapye river to its source and then went on to the Macloutsie. Trading in ivory and skins, together with a companion named Millen, in 1832 he visited Shoshong, he and Millen being the first recorded white men to reach so far north.

Botswana ivory was being exported from present-day Lobatse in 1780, and probably east and west from regions farther north as early as the seventeenth century to meet Portuguese demand on the coasts. But this was traded between African tribes and Europeans were not involved. By 1840 the Afrikaners, moving up from the Cape in search of new lands to settle, had practically occupied the Transvaal and Orange Free State, embarking upon a relentless slaughter of the game. Ten years later they began to commercialise on the skins and ivory, and as the supply rapidly diminished traders pushed westwards to Botswana where they bartered with tribes on the border for these items. At first they had all that they wanted without venturing into the interior, something the tribes were reluctant to allow. J. H. Wilson, of unknown origins, was already established at Kolobeng in 1847, and the following year visited Shoshong. Others soon followed, but near to Serowe was the limit of discovery as beyond was said to be a waterless waste.

1.2 Peoples of Botswana

Covering almost 582,000 km^2, Botswana is a little larger than France. In Livingstone's day it was estimated to have only some 40,000 inhabitants, most of these concentrated on the south-east border close to the Molopo, Notwane and Limpopo rivers. They comprised four main tribes: the Kwena (the crocodile people), Ngwato and Ngwaketse (all groups of the Tswana); and the Kalanga. Gaborone, the present-day capital, takes its name from the chief Gaberone who headed a small group of refugee Makatese peoples from the Transvaal, squatting in Kwena territory. The Kgalagadi lived to the south-west, a dispossessed tribe pushed into the more inhospitable areas of the fringes of the Kalahari desert; whilst in the far north-west near

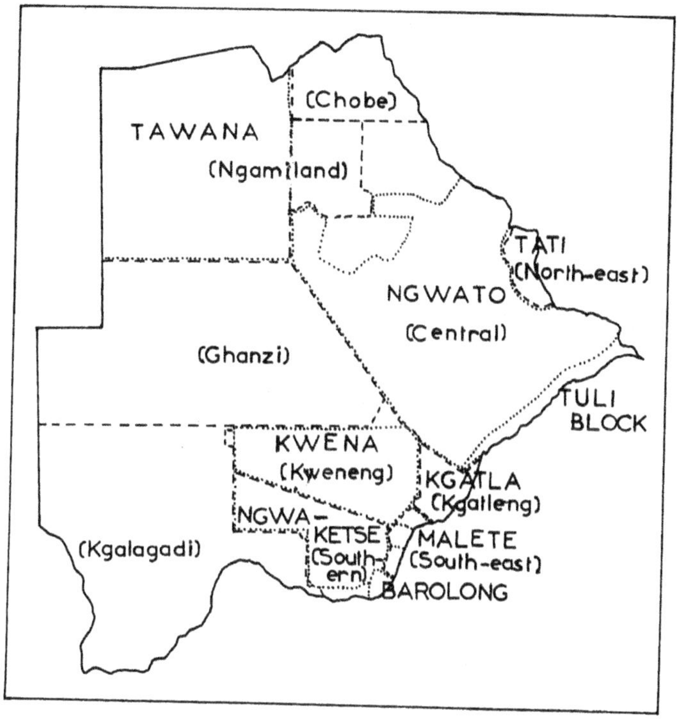

Fig. 1.1 Map of Botswana showing former tribal areas and modern district names

the Okavango swamps were the Tawana. A number of smaller tribes also existed in the north.

Each main tribe had a chief and a well-developed system of customary government. Most notable in Livingstone's day was Sekgoma I, chief of the Ngwato, the most powerful of the tribes. Later the Ngwato were ruled by Khama III, regarded as one of Africa's greatest chiefs.

Inhabiting the vast remote desert hinterland were the San or bushmen. Little yellow men living-out a stone-age existence at one with nature, able, in all but the harshest of droughts, to find underground supplies of water and existing on the roots and water storage organs of the desert plants, hunting the wild game with poison arrows (Fig. 1.1).

1.3 The Waterless Country

Livingstone provided a remarkably concise and apt description of the country in 1857: "The space from the Orange River in the south, lat. 29°, to Lake Ngami in the north, and from about 24° east long. to near the west coast, has been called a desert simply because it contains no running water, and very little water in wells. It is by no means destitute of vegetation and inhabitants, for it is covered with grass and a great variety of creeping plants; besides which there are large patches of bushes and even trees. It is remarkably flat, but intersected in different parts by the beds of ancient rivers; and prodigious herds of certain antelopes, which require little or no water, roam over the trackless plains. The inhabitants, Bushmen and Bakalahari, prey on the game and on the countless rodentia and small species of the feline race, which subsist on these. In general the soil is light-coloured soft sand, nearly pure silica. The beds of the ancient rivers contain much alluvial soil; and as that is baked hard by the burning sun, rain-water stands in pools in some of them for several months in the year."

Botswana forms a part of the Southern African plateau, an almost entirely flat and featureless land-locked area 950 m above sea level. The majority, except for the eastern border, is covered by the Kalahari desert or sandveld; 435,600 km^2, or 75% of the country. Deep, loose, sand-the Kalahari sands-which stretch from the Orange River south of Botswana where Harris and Cumming first met the great herds of game, in a broad swathe right up through eastern Angola and western Zambia, through western Zaïre and into Congo Brazzaville. It forms the largest expanse of sand in the world, covering 1,630,000 km^2, and in some places in Botswana forming a mantle over 300 m deep. In Zaïre tropical rain forest grows upon it, but at its southern extreme in Botswana, with only a scanty rainfall prone to regular drought, it is in places a true sandy desert. The Sahara of the south.

Except for a narrow limestone outcrop in the extreme west at Ghanzi, only in the east are there firm soils and rocky outcrops, dissected by once-permanent rivers and streams. But in the north-west is the vast 15,000 km^2 inland delta of the Okavango, fed by the Okavango River from the southern highlands of Angola spilling the whole of its waters into the sand. And to its east the equally vast dried-up saline lake-bed of the Makgadikgadi, a smooth expanse of glaring white salt 14,000 km^2 in extent. Once the waters of the Okavango reached here, overflowing from the swamps along the Boteti River, the anomalous river which flows in the dry season, several months after rain has fallen in the Angola highlands. In years of heavy rainfall it still flows as far as Mopipi. Reaching the limit of its forward impetus it may then flow backwards to the swamps, drying out until the next season except for a few pools in its bed. The closer to the swamps the more water remains, so it is much wetter in the Makalamabedi region than higher up towards Mopipi.

But waterless is the word characterizing Botswana, for almost the whole of its interior is without perennial surface water. But it was not always so. At one time the Kalahari was wetter as networks of fossil rivers with their dry, white salty beds, testify. Deep in the south, the dry Nosop River, a former tributary of the Molopo, is flanked by stabilized sand dunes. Ridges in serried parallel ranks covered with sparse tussocks of vegetation. They may not be of any great age, perhaps created in the late eighteenth century when conditions became much drier.

At this time the Nosop flowed in the wet season and there were large numbers of Hottentots living along it, tilling the land and running large herds of stock. As conditions became drier this could have led to overgrazing and the formation of the dunes, forcing the inhabitants to evacuate the area and permitting vegetation to cover the dunes. Not far distant, at Bokspits, coloured farmers have created immense sand dunes in the space of thirty years with their uncontrolled ravenous sheep and goats.

1.4 From Frost to Drought

Rain falls mainly between October and April, but in the southern extremity of the country it averages below 250 mm a year and is variable in the extreme, added to which Botswana falls within that part of southern Africa prone to an average 20-year drought cycle. For ten years the trend in rainfall is above average, followed by ten years below. In the extreme north-east, along the Chobe River, rainfall reaches as high as 650 mm per year, but the vast hinterland receives only half of this, between 300 and 400 mm. Summer temperatures can reach 42 °C leading to high evaporative water loss and soil temperatures of 70 °C. During winter temperatures can fall dramatically, for in May to July cold winds blow up from the Antarctic and night-time temperatures can drop to well below freezing. Along the Nosop River a temperature as low as − 18 °C has been recorded, and − 10 °C is not unusual.

Stepping out of your tent in the morning in July to find it covered with hoar frost bristling an inch thick and the water frozen into ice in the wash bowl, it is hard to imagine that you are in Africa, land of torrid heat and tropical beasts. But by nine-thirty the sun comes up, the frost is dispelled as if by magic, and the animals quickly warm up in its comforting rays. The trouble comes when the cold winds bring cloud and drizzle, then, with the sun blocked out, unable to keep warm many animals sicken and die of pneumonia. Especially those that have laid down insufficient fat reserves.

In 1854 there was a very severe snowstorm in the Shoshong area, and in July 1964 there was a light fall at Tsabong which persisted for some hours, but normally snow does not fall north of the latitude of Johannesburg. Not far south of Johannesburg, in March 1849, mistaking it for a lion Chapman fired at the carcase of a wildebeest which had been killed by snow.

The farther north one travels, the more hospitable does the land become for wild life, until one reaches the tropical Chobe River and the inland swamps of the Okavango Delta, where night temperatures never fall as low as they do farther south. But there is a price to be paid for living here, in the form of mosquitoes and malaria, tsetse fly and trypanosomiasis, as well as other parasites and diseases absent from the drier regions.

1.5 Chobe in the Past

The visitor entering Chobe National Park from Kasane today would recognize Selous's 1872 description of the vegetation: "a dense continuous jungle, interspersed with large forest trees, came down in most parts to the water. This jungle-covered land rises in some places abruptly, in others in a gentle slope, leaving along the shore a margin of open ground (from ten to a hundred yards broad), covered with short grass."

Buffaloes were numerous and on Puku Flats one morning, besides a black rhinoceros cow with tiny calf, he saw kudus, zebras, "and as usual any amount of impalas, pookoos, and baboons, which latter quaint-looking beasts swarm along the Zambesi and Chobe".

The only species you would not encounter today is the black rhinoceros. His observation on the number of baboons occurring is of interest because it is commonly alleged today that where they are numerous this is because their natural predator, the leopard, has been reduced.

Selous considered impalas were more plentiful along the Chobe River than anywhere else, herds of up to 100 being encountered. Met with along all of the rivers from the Notwane northwards, in the rainy season they could be found at places like Boatlaname in the east of Kweneng District. A curious fact today concerning this species is that from time to time an isolated individual allegedly turns up on the Nosop River, curious because the nearest known population is in the area south of Lobatse about 500 km away (Plate 1.1).

Plate 1.1 Selous found impala numerous along the Chobe River in 1874

1.6 The Kalahari Desert

As far as nutrients are concerned the soils are poor and the sands even more so, yet given rainfall, abundant green vegetation covers the Kalahari desert, perhaps only because it is ephemeral in occurrence and there is not a persistent uptake of the limited supply of nutrients. Livingstone, who amongst his many other accomplishments showed himself to be an observant ecologist, wrote: "The quantity of grass which grows on this remarkable region is astonishing.... It usually rises in tufts with bare spaces between, or the intervals are occupied by creeping plants which, having their roots buried far beneath the soil, feel little the effects of the scorching sun. The number of those which have tuberous roots is very great... Leroshua, is a blessing to the inhabitants of the Desert... on digging down a foot or eighteen inches beneath, we come to a tuber, often as large as the head of a young child; when the rind is removed, we find it to be a mass of cellular tissue, filled with fluid much like that in a young turnip... Another kind, named Mokuri,... deposits underground a number of tubers some as large as a man's head... But the most surprising plant of the Desert is the 'Kengwe or Keme' (*Cucumis affer*), the water-melon. (*Citrullus lanatus*-tsama melon, he is confusing it with the perennial desert cucumber *Cucumis africanus*, but also an important source of moisture). In years when more than the usual quantity of rain falls, vast tracts of the country are literally covered with these melons; this was the case annually when the fall of rain was greater than it is now... It happens commonly once every ten or eleven years, and for the last three times its occurrence has coincided with an extraordinary wet season. Then animals of every sort and name, including man, rejoice in the rich supply. The elephant, true lord of the forest, revels in the fruit, and so do the different species of rhinoceros... The various kinds of antelope feed on them with equal avidity, and lions, hyaenas, jackals and mice, all seem to know and appreciate the common blessing". In his private journal he added wild dogs also ate them.

Gilarmi Farini, real name William Hunt, the American entertainer who claimed to have discovered a lost city in the desert, wrote in 1886: "How could the Kalahari ever have been called a desert with so much vegetation, such countless acres of grass, reaching now nearly to the horses' backs?". He pointed out that even when there had been no rain for a year the grass was equal to ordinary hay, while in the dry season the tsama melons remained good for a year.

The San once believed the tsama melons were sweet or bitter according to whether they happened to be deposited in the dung of white or black rhinoceroses, or of the elephant. Nowadays only bitter melons are found (Plate 1.2).

With the rain numerous shallow pans fill with water, the desert springs into life, and the game multiplies. There is not only grass, along the edges of the ancient now dry river beds and in the valleys between the sandy hills, the roots of acacia trees find moisture and parts of the Kalahari are densely covered with woodland. Because it is not a desert in the sense of being devoid of vegetation, the Kalahari has been termed a "thirst land", a name first coined by William Baldwin hunting in Botswana in 1859, and taken up by Parker Gillmore in 1876.

Plate 1.2 The Kalahari desert

On the route from Lephephe to Shoshong Gillmore described it as "Sand, sand, sand, so deep that the felloes of the wheels are entirely covered, the brush short and scant, the few trees very dwarfed, and the surface of the soil rolling-such are the characteristics of the country. No game, no birds, naught cheers the scene-it is a land of desolation, of waste, that nothing can ever be done to improve... Such is the south-eastern end of the Kalahari, the Great Thirst Land."

1.7 Flora and Fauna

The flora and fauna of the Kalahari, of which Botswana is mainly comprised, are to some extent transitional between the flora and fauna of neighbouring regions; the Karoo-Namib desert region to the west and south-west, the Zambezian region to the north-east, and to the east and south-east the Southern Savanna region. Floristically Botswana shows greatest affinity to the Zambezian region. In contrast to this its fauna shows greater affinity with the South West Arid region.

When invertebrates, fish, reptiles, amphibians, birds and small mammals are taken into consideration, faunal diversity is relatively limited; but of the 101 mammals recorded, the 40 "large" mammal species and seven possible races, represent a relatively rich large mammal fauna. Although Botswana has two types of large mammal fauna, separation appears to have become more distinct within the past century due perhaps to increasing aridity compounded by human destruction of water-dependant species. In the driest southern part the fauna is that of the South West Arid Biome, characterized by a few water independent species: namely gemsbok, hartebeest and springbok. Species shared with the wetter Zambezian region, indistinguishable from the East African Sub-region, are blue wildebeest, greater kudu, eland, giraffe, and ostrich.

1.8 The Abundance of Game

Despite the criticisms levelled at him on account of his propensity for wilful slaughter, there has perhaps been no more enthusiastic pen describing the game of Botswana than that of Cumming. It began at the Molopo River, a now dry watercourse, which in 1844 was to Cumming his "darling little Molopo". Large herds of buffaloes then inhabited its reedbeds and those of the Notwane, but even had they not been remorselessly slaughtered by everyone who was to subsequently pass that way, by 1907 Selous was to report there was no longer water enough in the Molopo to support a herd. In August 1826, just east of Bray, Bain had found only a filthy pool of water in the bed of the river which he thought could not last for more than two days. Orpen recorded at the end of June 1852 that the river near Bray was a sandy bed and he had to dig pits for water, but inland to the north he met giraffe, eland, gensbok, buffalo and other animals, except for rhinoceros. Large herds of game had been moving about near Khakhea.

Cumming shot all of his game, a staggering quantity, never farther north than Lose, near Shoshong, including elephant. At the end of May 1844 he reached Mabotsa in the western Transvaal, where he was received by Livingstone. The position of Mabotsa is uncertain, even Livingstone himself was not sure exactly where it was, but it appears to have been on a branch of the Marico about 15 or 35 km east of Lobatse. Botswana was probably entered through Pioneer Gate and the first halting-place, on a branch of the Notwane, was reached in one day. With fresh trek-oxen travellers may well have covered 20 km, although half this was the daily average of an ox-waggon.

Before reaching this spot Cumming crossed the headwaters of the Molopo where reedbucks were abundant as well as lions. He also found reedbucks at the junction of the Marico with the Limpopo but not along the Limpopo River itself. Although the Molopo was "a darling little river", in July 1846 it had dried into muddy pools.

Arriving at the Molopo in 1845, Oswell saw seven different kinds of animal within view; zebras and buffaloes in large herds, together with scatterings of springboks, hartebeests and wildebeests. Altogether he estimated he could see some 3000 animals.

Livingstone was already in Kgatla country, north of the present-day capital of Gaborone in country which Oswell described as "the well-wooded and watered valley… a broken country full of game… a hunter's paradise. For the first time tracks of rhinoceros, giraffe and other unknown creatures were abundant." Giraffes were seen every day in herds of five to 30, and buffaloes were numerous.

Both Cumming and the less-known hunter Methuen, were hunting in eastern Botswana during a period of drought and the quantity of game which they recorded, always in the vicinity of waterholes and water courses, may have been exceptional, game moving eastwards from a much wider area out of the central Kalahari concentrating upon the remaining water.

But the eastern border of Botswana was also once clothed with woodland rich in game because it represented a vast "ecotone", a region where two types of habitat meet, tending to produce greater variety than either habitat alone. It formed an ecotone rich in species which exploited the Kalahari desert during the wet season, retreating into the hard pan woodland in the dry. Chapman wrote in 1868 from his fifteen years' of experience: "During the greater part of the year it [the desert] is overrun by thousands of elephants, giraffes, rhinoceroses, elands, and other animals who stay in it until driven away by the drying-up of the waters: but no sooner do the clouds begin to gather than they turn their faces thither again, lingering on its borders until the rain falls on its loose and thirsty sands, when they enter triumphantly, and, dispersing in every direction, soon fill it with life."

But in May 1852, on his first visit down the Notwane he had climbed a small hill somewhere in the region of Mookane and, "had a look at the vast and dismal scene around me."

Of the region south of Kanye, the traveller Andrew Anderson, who had formerly been a magistrate at the Cape, wrote that in 1869 the plains swarmed with game, including lions. And of the area west of the Notwane River, "It was, when I first knew it [1869], full of game, but since the natives have obtained rifles, they have greatly reduced it." The northern region of the Molopo River he referred to as "a complete forest, fine trees, bush and open glades." In the region of Mochudi the country was full of lions, so bold they came close to the town: "They follow the game, more particularly quaggas or zebras, and, as there were plenty of them, also koodoos, hartebeest and wildebeest" (Plate 1.3).

On the road to Shoshong, about 1869, the "whole country swarmed with game, lions and wolves." By "wolves" he meant the brown hyaena *Hyaena brunnea*. There were hartebeests, impalas and zebras. "The ostrich was then plentiful." Ten years later south of the confluent of the Notwane with the Limpopo, on the Limpopo river, Parker Gillmore wrote: "Scarcely ever have I seen game in such abundance as in this part of the country, spoor of every description being discernible in all directions."

In 1845 Oswell did not find elephants until he reached Kaa country, the area of Shoshong, although he once hunted them in Kwena, near Gaborone. In 1835 the Kaa lived three days from the Ngwato in the Shoshong Hills because they were enemies of one another. This created a no-man's land in which game could roam relatively unmolested.

But game it seems was plentiful everywhere, even in the extreme arid west. Just over the border from Ghanzi at Rietfontein on the upper Hanahai, in 1851 Galton found an abundance: "we were, without any doubt, in a game-country. The river-bed was trodden like the ground in a cattle fair by animals of all descriptions. The water lay in pools among the rocks." Elephants had come "in droves" to drink.

Plate 1.3 A lion in the Chobe park

1.9 Species Always Rare

At the limits of their ranges, those species uncommon to rare over a century and a half ago, have remained so. These are the oribi *Ourebia ourebi rutila* and the Sharpe's steenbok *Raphicerus melanotis sharpei*. The oribi occurs only in the north-east in the Kazuma area on the black clay plains, and in the Linyanti-Kwando area; its habitat threatened by farming schemes. Holub recorded it from Makgadikgadi but he probably mistook a steenbok for it. Selous recorded it in the Kazuma valley "an open boggy flat of only a few hundred acres in extent."

The Sharpe's steenbok is also limited to the extreme north-east in the Pandamatenga area, near Kasane and formerly near Kachikau. One was shot by Baldwin in 1860 north of Pandamatenga.

The Vaal rhebok *Pelea capreolus*, reported to occur in the hills of eastern Botswana in the region of Lobatse to Gaborone by Bryden and others, has never been positively identified there and possibly the original references referred to the mountain reedbuck which occurs in these hills (one was found caught in a fence near the Gaborone Dam in 1989). Methuen recorded that the first animal he "saw and wounded" in Botswana was a grey rhebuck, which he stated was well known in Cape Colony. Perhaps he had mistaken a grey duiker *Sylvicapra grimmia* for it, although H. Neale stated it was "fairly plentiful" in the hills near Gaborone in 1930.

1.10 Other Scarcer Species

Bushbuck *Tragelaphus scriptus*

The Chobe bushbuck *Tragelaphus scriptus ornatus*, found from the Boteti River northwards, is simply a variant differing in size and colour from those found further north and south. The common bushbuck varies widely in colour and pattern throughout Africa, even within a single population in overlap areas.

The trader Fredrick Green claimed to have been the first to obtain a specimen of the Chobe race, on the Thamalakane River in 1852, "I believe I was the first European who.... shot a specimen of this buck." The skin was destroyed but he published a description from his journal entry in the Grahamstown *Eastern Province Monthly Magazine* in 1856. In 1855 he obtained another on the Thaoge River. Chapman records shooting an antelope in June 1854 at the junction of the Thamalakane and Boteti Rivers answering the description of the Chobe bushbuck. It was "the loveliest animal that, it seemed to me, was ever created", he wrote. He presented the skin to Moffat, "who intended sending it to the British Museum". Writing of this in 1862 Baines states that a new antelope "was once killed by Chapman, who, after carrying the skin about for a long time, and finding no evidence of getting nearer the colony sent it to Mr. Moffat, to be forwarded to the British Museum. The chances of transit are, however, so precarious that, with the best intentions of all concerned, it may very likely have miscarried."

The British Museum did, however, in 1860 purchase from Verreaux, one of two French brother collectors, a skull with horns and skin allegedly collected by Chapman in the "Zambesi Valley", and said to be that of the sitatunga. Mr. Moffat's trust appears to have been misplaced!

But the specimens from which the race was described were sold to the British Museum by Selous in 1881. Frederick Barber donated another specimen in 1899. Selous correctly believed it to be one and the same species with other bushbucks. In his time in certain districts along the south bank of the Chobe it was more common than anywhere else, but found only where dense bush comes down to the water's edge.

On the Limpopo the adult bucks are a brownish grey, often lacking any sign of spots; and the adult does dark russet red with a few white spots. Cumming described the Limpopo animal as a new species, the *Antelopus roualeynei* or Limpopo bushbuck (now *Tragelaphus scriptus roualeyni*).

In the Chobe bushbuck the adult buck is a very dark russet red, in places merging into a deep brownish black and with many large white spots along its lower flank, coupled with in some cases as many as eight well-defined white vertical stripes. A mane of white hair runs the length of the back from shoulder to tail. The adult doe is pale russet, spotted with white and with three or four white stripes on the flanks and a deep black line running the length of the back. On the Limpopo the young are more spotted than the adults, but in the Chobe race this is reversed, and the horns of the buck are generally shorter than those found on the Limpopo. Selous, who

Plate 1.4 Chobe bushbuck ram. A race of limited distribution

shot specimens of the Chobe bushbuck in 1877 near Linyanti, considered it formed a connecting link between the dark-coloured bushbuck of Cape Colony and the smaller harnessed antelope of north-western Africa, which has both vertical and horizontal white stripes.

Concern has been expressed that the Chobe bushbuck's existence may be threatened by the removal of the understorey in its habitat by elephants, but the bushbuck is one of the more tenacious of antelopes and usually maintains an existence where other species have disappeared and so far it has maintained a hold on the Chobe waterfront (Plates 1.4 and 1.5).

Sitatunga *Tragelaphus spekii*

Rarely seen, the bushbuck's close relative, the sitatunga *Tragelaphus spekii*, is not uncommon in the swamps of the Okavango and the Linyanti. Named after Captain Speke who collected horns of it in Tanganyika in 1861, this marsh-loving antelope with its elongated hooves had been reported long before in Botswana as Speke was aware. Probably known to Livingstone in 1849, as Oswell obtained a cloak made of parts of its skin, Leyland obtained the horns and skin of one in 1851 taken by a local on the Thamalakane River, which he called the *Antelope Anderessoni* in recognition of Andersson's first description of it.

Andersson had described the animal, including its long hooves, in his book *Lake Ngami* published in 1856 (Leyland's book was not published until 1866), but his best materials were left behind in Africa. He claimed to have seen a pair only at a distance but was aware of the nature of the hooves from local descriptions, the people frequently catching the animal in pitfalls and by spearing. He examined some skulls,

Plate 1.5 Chobe bushbuck doe

as well as the cloak brought from Lake Ngami by Oswell, and sent an "imperfect" skin to the British Museum, which Dr. Gray thought to be that of a type of bongo.

The British Museum does not list Andersson's "imperfect" skin but does list two sets of earlier specimens than Speke's. These were from Lake Ngami presented by J. A. Green in 1856, apparently an error for his younger brother C. A. Green who was at Lake Ngami in 1855. For some reason they did not receive any attention.

When Chapman shot the antelope in June 1854 at the junction of the Thamalakane with the Boteti he made no mention of any peculiarity regarding the hooves; and when with Chapman at almost the same spot in April 1862 Baines wrote the sitatunga "has, it is believed, never been shot by a white man". So the specimen sold to the British Museum by Verreaux in 1860 as being that of a sitatunga shot by Chapman, was probably a Chobe bushbuck.

But in 1906, John Kenny, who had been Chapman's servant on the expedition to Lake Ngami in 1859, presented two pairs of sitatunga horns to the British Museum. So the animal must have been familiar to Chapman and his party at that time although Chapman himself makes no specific mention of it. It was not until 1881 that Selous presented to the British Museum some horns and feet taken on the upper Boteti. As a result the northern Botswana animal was designated a separate race *T. s. selousi.*

In 1900, Reid, a South African sportsman, donated two doe skulls from the "Left bank of the Linyanti Valley", wrongly located by the Museum as in British East Africa.

J. A. Nicolls, a South African resident who visited Lake Ngami with John Strombom in mid-1889, was informed that sitatungas were very numerous on the Thaoge. Giving permission to hunt them, the Chief Moremi exclaimed that there must be some big medicine in the naakong's belly which white men were very anxious to get, *naakong* being the Tawana name for sitatunga. Nicolls found them on a papyrus island which was choking the bed of the Thaoge, and missing a buck shot a doe found to be accompanied by a doe calf of about ten days old, which was captured. Moremi generously donated a cow and two goats to supply milk for it and surprisingly it was brought back alive to Mafeking in a waggon. Nicolls donated it to the London Zoo, where it arrived at the end of 1890 and lived until 1905.

Although easily speared from canoes, especially in the dry season when the papyrus may be burnt off, an estimated 4000 existed in the Okavango in 1990, one of the biggest concentrations on the continent. They have declined along the Kwando-Linyanti rivers due to poaching and have been heavily poached along the Chobe River near Kasane, but 17 were seen on Sedudu Island in 1990 (Plate 1.6).

Puku *Kobus vardoni*

Also to suffer decline has been the rare puku, which Selous found so relatively numerous along the Chobe waterfront. In 1990 these were down to 120, although the

Plate 1.6 Sitatunga. The first photograph and capture of a sitatunga doe in 1890, wrongly labelled a 'waterbuck'. From Wills, W. A. and Collinridge, 1894. The Dowfall of Lobengula. London, "The African review" offices

maximum number seen in 1972 was 43. Selous, who first shot pukus here in 1874, referred to them as "a rare species of water-buck". Later, "At the point where we struck the river on our way back... stretched a large flat piece of ground, in some parts over half a mile broad, lying between the steep, forest-covered, jungly sand-belt and the bank of the river. This flat might be from six to eight miles long, and lay in the form of a semi-circle... The greater part of this extensive tract... was open, though here and there patches of bush were scattered over its surface, and near the river grew many very fine wide-branching camel-thorn trees... The number of pookoo on these flats quite surprised me. Sometimes troops of more than fifty of them were to be seen together... Owing to the great numbers of these antelopes, I christened this place the "Pookoo Flats".

It retains this name today. But after the assassination of a local chief on the Zambian side in 1876, a number of people fled to the south bank of the Chobe and committed great havoc among the herds, so that when Selous again visited the area in 1877 he never saw more than ten or a dozen in a herd, "and not one for every ten I had seen there in 1874."

In Botswana, Selous only ever found the puku in a small tract of country extending along the southern bank of the Chobe for about 100 km westwards from Kazungula. Further west he was to write: "Singular to say, the pookoo antelopes, which were so abundant some thirty miles farther back, had now entirely disappeared, though the character of the country and the vegetation remained precisely the same in every particular. I cannot think of any reason to account for the curious localisation of this species."

Two South African travellers, Dr. Aurel Schulz and August Hammar, having travelled up the Chobe River along the south bank, reported it in 1885 as appearing to be rare, meeting some only about the region of Kavimba.

It is never found more than 200 to 300 m from the river and is usually seen cropping the short grass along the water's edge or lying in the shade of the bushes. Those on the north bank have now been completely eradicated.

There has always existed some confusion as to where the first or "type" specimen was seen. Livingstone illustrated in his book Missionary Travels a new antelope introduced to science by Oswell and himself, and mentioned seeing it well north in Zambia. But he first recorded seeing it in his journal somewhere in the Caprivi Strip between the Chobe and Zambezi along the north bank of the Lilonga, which flows into Lake Liambezi, "We saw a number of antelopes called poku, very much like the reed buck but probably different" he wrote in his journal on August 4th 1851. Livingstone named it himself after Oswell's friend, Major Vardon. As the first sighting was not published, the type locality is given as 50 to 60 km north of Libonda on the Zambezi River.

Selous sold a specimen to the British Museum in 1881 which was listed as coming from "Umparvia, Chobi valley"; a misprint for Mparara, an island in the Chobe River near Kazungula (Plates 1.7 and 1.8).

Waterbuck *Kobus ellipsiprymnus*

Plate 1.7 A puku ram. Still where Selous found them in 1874, only a handful remain today

Plate 1.8 Puku does

Plate 1.9 Waterbuck on the Chobe

For some reason its cousin the common waterbuck has never done well along the Chobe. Numbering only some 50 about 1990, Selous found it was met with sparingly on the Chobe and the Savuti. He found none on the Boteti and Mababe rivers in 1879, although Nicolls shot a large buck in the region of Makalamabedi on the Boteti in 1889 and also saw a herd of does there. But it does better along the Limpopo where good numbers still exist on the adjacent farmland. Cumming saw his first near the junction of the Notwane and the Matlabas rivers, and Methuen reported them as becoming "very common" downstream on the Marico River (Plate 1.9).

Klipsringer *Oreotragus oreotragus*

Seldom spotted today, Leyland recorded the klipspringer as very timid and shy and Holub recorded the "Bakwena heights" as being its home. Bryden in 1890 recorded them in the hills near Serowe. It was a Tswana tradition that the klipspringer could attract rain, and in the drought of 1889 the Bangwaketse organized live hunts for them. The captured animals were then carried about in their arms, while they pinched them to make them squeal in the mistaken belief that their plaintive cries would bring rain.

Bryden considered the klipspringer "abundant" throughout the interior wherever rocky hills are found but these are limited to eastern Botswana. Nicolls and Eglington stated it to be "plentiful" in 1892 in the hills around Kanye and Molepolole. Two races occur with a clear break between them; the Transvaal race in the south up to Gaborone changing abruptly to Stevenson's race from Dibete to the Tati area.

In 1892 a storekeeper at Kanye had several tame ones which, inquisitively nibbling at everything, all died from ingesting the cast off lead tops of whisky bottles.

Sable *Hippotragus niger* and Roan *Hippotragus equinus*

Of the larger less frequently met species, the sable and the roan antelopes were just as uncommon a century or more ago as they are today. Probably north-west of Otse, Methuen in 1845 saw a herd of 15 to 20 sable, "we never had another chance of obtaining this magnificent antelope" he commented, so it was not that common in the area. Baldwin reported shooting them as well as roan antelopes, elands, zebras and warthogs, at Jare Pan in 1860; but Selous saw his first sable antelope, "one of the handsomest animals in the world", at the Ramokgwebana river in 1872.

"All along the Chobe River, as far as I have been [the outfall of the Savuti], I have met with this antelope, though sparingly. In the Mababe country, and on the road leading from there to the Bamangwato, I neither saw a sable antelope nor the spoor of one." Bryden reported in the early 1890s that there were still a few to be found in the Ramokgwebana area, and that they were found in moderate numbers around the Chobe River.

Cumming had found them in the Tswapong Hills, only some ten years after it had been first made known to science by Harris in 1838 from a specimen collected in the Magaliesberg near Pretoria, where it is now extinct.

Roan antelopes were seen by Cumming north of Molepolole, and in Selous's time a few occurred on the Notwane and the Limpopo rivers. It was not uncommon near Tati, but found only sparingly along the Chobe and in the Mababe area. Nowhere found in large numbers, a herd of 20 was rare then, just as it is today (Plate 1.10).

Tsessebe *Damaliscus lunatus*

Chapman was informed in the early 1850s that the Mababe Flats swarmed with game of every description in great multitudes, including rhinoceroses, buffaloes, elands, tsessebes and others. At the end of the dry season large herds of tsessebes congregated together here. They were not found south of the Boteti River although

Plate 1.10 Sable antelope bulls

Bryden believed, erroneously so, they were formerly "very abundant throughout Bechuanaland". Now [1890] they were seldom met with and he never even saw the spoor of one. But in 1869 Baines wrote he had never before seen a tsessebe until in June of that year he saw one on the Ramokgwebana River. Ten years later Gillmore claimed to have come across a dozen tsessebe somewhere near Shoshong. He comments they were at one time found in great numbers when they were captured in *hopos* by the Tswana, and he attributed the lessening of their numbers to this capture.

The tsessebe appears to give way in competition with cattle, but the Deka flats north of Pandamatenga are said to have held one of the largest concentrations known anywhere in Africa prior to 1950, after which year many thousands were shot by the Colonial Development Corporation to make way for cattle. Before the massacre herds ran into hundreds of head, but by 1960 rarely exceeded 25. However G. C. Shortridge reported that in the Okavango valley at the beginning of the 1930s, the herds never exceeded a dozen and the most common size was three to six, although temporary concentrations occurred at certain seasons. In 1990 their total numbers in the country were put at over 7400, with most of these in the Okavango Delta; but their numbers could have been as high as 12,000.

Hyrax *Procavia capensis*

The little Cape hyrax, not a "big game" animal but a bizarre relative of the elephant, is probably a relatively uncommon species because its pelt has always been sought after. In the hills around Kolobeng in 1850 Leyland reported that they occurred in small groups of three or four "instead of eighteen or twenty as in the Colony." In 1873 it was found in rocks along the Metsemotlhaba valley near Mosopa, but even then Holub recorded that "on account of being so continually hunted by the natives, [they] are very shy." The people often kept them, perhaps to eat rather than as pets, tied up by a back leg with a piece of cord which the hyrax will not attempt to gnaw through.

1.11 The Fabled Lake

At an early period of the nineteenth century rumours had reached Europeans of a vast lake in the interior of southern Africa. In 1813 the missionary John Campbell when at Kuruman was informed of "a nation called Mampoor, who reside on a great water, across which they could observe no land, and on which they observed the sun to set." In 1834 the Cape Colony government sent an expedition to find it led by Dr. Andrew Smith. Smith turned back in September 1835 about 50 km south of Mahalapye as he was told that there was no water ahead for three days, the locals making the journey only in the wet season. Several parties of coloureds had tried but had always turned back for want of water.

At least it is thought that none went further. The route lay across the great Ntwetwe salt pan and on its farther shore stands a huge baobab by the side of Gutsha pan, a

Plate 1.11 1859 signature of one of the earliest explorers still to be seen carved on a baobab tree at Gutsha pan

pan which in those days held water and was where travellers camped. Several were to carve their initials on this tree, and still to be seen today are those of H. Van Zyl, 1851, a disgruntled Afrikaner politician who settled at Ghanzi and became an elephant hunter; and Frederick Green, 1859. But most intriguing is a fading trace of a date without initials of 1771 (Plate 1.11).

Then to Botswana came the missionary Rogers Edwards accompanied by the young David Livingstone, who made an exploratory journey from Kuruman the farthest north mission station in southern Africa, as far as the Bamangwato in 1841. Having read Campbell's account of the great inland sea of water in the midst of the desert, Livingstone wrote to his sister in February 1841 that he would like to be the "first" to see it. But it was not until 1849, accompanied by William Cotton Oswell, an English gentleman seeking adventure; Murray, Oswell's hunting friend, and Wilson a trader; they, the first recorded white men, reached Lake Ngami.

The way to the heart of Botswana had been opened up.

Chapter 2
Impressions Good and Bad

Abstract Early hunters' descriptions of the abundance or otherwise of animals, Cummimg's descriptions of game, "perhaps more [game] than any other district throughout the world". Livingstone's descriptions of the countryside. Anderson, Bryden and others. Changes. Selous and Boteti River changes. Conflicting impressions, negative impressions of the arid countryside.

Keywords Big game · Botswana · History · Quotations

2.1 Cumming's Country

Cumming proposed travelling north-east over the border in Botswana some 320 km to Ngwato territory to hunt for elephants said to be abundant there, although it was put to him he would find them in the Kwena country, 80 km to the north of the Otse region. His descriptions of the country and its game may wax somewhat lyrical, but are full of vigour. The buffalo's boss reminded him of the rugged trunk of an oak tree and the horns "descending, and spreading out horizontally, completely overshadowed the animal's eyes, imparting to him a look the most ferocious and sinister that can be imagined."

To the north of Mannyelanong Hill he came across two white rhinoceroses, but as he continued northwards we would not recognize in this dry parched, treeless country of today the rich verdure of yesteryear: "through a country of increasing loveliness. Beautifully wooded hills and valleys... stretched away on every side, with rivulets of crystal water in the valleys and the spoor of large game very abundant... halting beside a stream of delicious water... halted at sundown beside the broad and sandy bed of a periodical river, through which ran a crystal stream".

Climbing the hills he exaggeratedly termed "a lofty mountain-range", he found large colonies of baboons and a few klipspringers, "we entered the gorge [the Kolobeng river or river of wild pigs] and wound along the margin of the stream which danced and sparkled down its abrupt and rocky channel, forming a pleasing succession of babbling streams and foaming waterfalls."

C. Spinage, *Big Game of Botswana*, https://doi.org/10.1007/978-3-031-60033-3_2

Kolobeng was named after the warthog which, according to Leyland in 1850, frequented the river "many years ago". Already in his time "rarely to be met with; I only saw one the whole time I was in the country [in South Africa]."

Cumming met buffaloes and lions. Emerging from a valley probably south of Molepolole: "We entered upon a more level country... Here water was very abundant. We crossed several streams and marshes whose margins were a mass of spoor of wild animals; that of rhinoceros, buffalo and camelopard [giraffe] being most abundant".

West of Mochudi at a waterhole near Boselkosi hill he saw 300 impalas, 12 tsessebes, 20 zebras and three mountain reedbucks. The next day 200 impalas, 50 wildebeests, 30 zebras, 30 tsessebes and two warthogs appeared; and also a solitary elephant. From the east to this waterhole he noted "an old established elephant and rhinoceros footpath". There were also buffaloes in the area and in the hills near Kopong he found klipspringer tracks. On the north side of the hills he found a waterhole where there was the fresh spoor of black and white rhinoceroses, buffaloes, wildebeests, tsessebes, kudus, klipspringers and other game. On the way he saw a troop of seven kudu does and also surprised three mountain reedbucks. North of Molepolole he had reported kudu as "tolerably abundant" (Plate 2.1).

Other game seen in the area were elands, roan antelopes, wild dogs and lions. But lions were everywhere. Conducting a night-long vigil by the waterhole a succession of buffaloes came to drink: "Towards daybreak we wounded a white rhinoceros, and soon after two black rhinoceroses fought beside us…" The next night three black rhinos came to the waterhole, two white rhinos and three other rhinos. Later, in 1848, he was to write of the area south of Kgale Hill near Gaborone: "It is a first-rate district for game when the country has not been ransacked by Griqua hunters." Already times were changing, and in 1845 he had written that the Notwane River "this year" was dry in October except at its source.

Plate 2.1 A kudu bull in the Chobe park

Travelling north he wrote: "I was much delighted with the little loch to which we then came [Sojwe pan?]; it covered about an acre, in shape a circle, and its margin was imprinted with the fresh spoor of a variety of wild animals, such as giraffe, rhinoceros, buffalo, sassaby, Pallah [impala], zebra, lion, etc."

"Here I found some old dung of elephants; and observed several full-grown trees torn up by the roots, and others that had been shivered".

He came across a black rhinoceros, a large herd of wildebeests, and zebras and springboks. South of Lephephe he met zebras, springboks, a troop of hartebeests, a large herd of wildebeests, impalas, and three black rhinoceroses.

Lephephe was an extensive pool of pure water with reeds on one side according to Cumming, but Oswell claimed the name meant "mouse", due to the water from the spring trickling a few yards and vanishing into a hole. When Oswell saw it in 1845 he considered it looked as if at one time the water had been much more extensive. He found lions here to be the worst he ever encountered for their audacity, on three occasions making determined attempts on his camp at night.

Reaching Masowe Pan south-west of the Shoshong Hills, Cumming christened it "the Elephants' own Fountain" because of the elephant spoor around it, the rock itself being worn down by their feet. A hundred well-trodden elephant paths led away from it in every direction "like the radii of a circle". Here he came across a pair of cautious roan antelope.

Of the Mahalapye (Mhalatswe) river where he spent some time hunting elephants, Cumming wrote: "This was one of the loveliest spots I had seen in Southern Africa: a bold bend of the river was adorned with groves of remarkably lofty and picturesque acacias."

Along the Limpopo, between Buffels and Martins Drifts, he found crocodiles, waterbucks, impalas, bushbucks and large numbers of hippopotamuses.

Steering for the mountain of "Guapa" [Tswapong Hills]: "I... soon started two pairs of ostriches... I next sent away a herd of zebras and brindled gnoos; next a sounder of wild boar; next a troop of giraffes; and, close to the vley where I intended to halt, I wounded a koodoo... I came to a black rhinoceros... I stalked in close upon a second black rhinoceros. A few minutes after firing at Borele [his name for the black rhinoceros] I was following a troop of pallahs close under the green rocky mountain, when, lo! a herd of lovely sable antelope caught my attention".

Later, "We fell in with immense herds of zebras and buffaloes numbering between three and five hundred in each herd, and towards evening with a numerous herd of elephants".

The farthest north that Cumming reached was Lose, then a three metre deep pool of good water in a small rocky valley forty km north-east of Shoshong. It still held deep water in 1875.

Writing of the area along the Limpopo east of Machaneng: "In the course of the day [17th November 1847] I saw the fresh spoor of about twenty varieties of large game, and most of the animals themselves, viz. elephant, black, white and long-horned rhinoceros, hippopotamus, camelopard, buffalo, blue wildebeest, zebra, waterbuck, sassayby [tsessebe], koodoo, pallah, springbok, seromootlooque [bushbuck], wild boar [warthog], duiker, steinbok, lion, leopard. This district of Africa contains a larger variety of game than any other in the whole of this vast tract of the globe, and perhaps more than any other district throughout the world; for besides the game which I have just noted, the following are not uncommon, viz. keilton, or two-horned black rhinoceros, eland, oryx, roan antelope, sable antelope, hartebeest, klipsringer, and grys steinbuck; the rietbuck [reedbuck] is also to be found, but not abundantly."

2.2 Like a Gentleman's Park

Almost a year before Livingstone's famous best-seller Missionary Travels and Researches in South Africa appeared, a correspondent in the London Journal December 1856 reported: "Europe had always heard that the central regions of southern Africa were burning solitudes, bleak and barren, heated by poisonous winds, infested by snakes and only roamed over by a few scattered tribes of untameable barbarians… But Dr. Livingstone found himself in a high country, full of fruit trees, abounding in shade, watered by a perfect network of rivers."

Livingstone allowed most rein to his eloquence perhaps when he described the country after leaving Goha Hills in the now Chobe National Park in 1853: "the valley on its northern side, named Kandehy, or Kandehai, is as picturesque a spot as is to be seen in this part of Africa. The open glade, surrounded by forest trees of various hues, had a little stream meandering in the centre. A herd of reddish-coloured antelopes (pallahs) stood on one side, near a large baobab, looking at us, and ready to run up the hill; while gnus, tsessebes, and zebras gazed in astonishment at the intruders… A large white rhinoceros came along the bottom of the valley… Several buffaloes, with their dark visages, stood under the trees on the opposite side to the pallahs".

In his private journal he wrote of Goha (which takes its name from the grub which the bushmen use for arrow poison): "It is covered over with dense forest, and in the North has a most beautiful valley, equal to that of any gentleman's park in England. The loveliness was increased by a White Rhinoceros sauntering across it and a gnu standing under a tree. A magnificent Boabob (*sic*) on one side added enchantment to the scene." Importantly there were no tsetse on the hill. The valley was about half a km to the north, but there is no valley within half a km; the Kasinka floodplain which was presumably Livingstone's valley is about 5 km distant.

On his way to the Goha Hills he passed through the complex of pans to the east of Savuti Marsh; "Pass Kingushe [at or near Tsotsoroga Pan] and several pools containing water. Spoor of elephants everywhere… come to Mopane trees with hard ground and much water… Much game about. Vines". Three days later he repeats: "Many vines. Much game. Bushmen everywhere, finely formed men…

"Passed through a lovely country of mopane trees chiefly. It is like a gentleman's park. Much water and game. Vines festoon the trees. The grapes are about equal to Bramble berries at home. Rhinoceros buffalo & elephants abound…

'As we went north the country became very lovely; many new trees appeared; the grass was green, and often higher than the waggons; the vines festooned the trees, among which appeared the real banian (*Ficus indica*), with its drop-shoots, and the wild date and Palmyra".

When they had been travelling along the Boteti river in 1849, his companion Oswell wrote: "The scenery generally along the river was magnificent. Trees of great size, rich in foliage, fringed it on either side; now it is shut in between high steep banks, and runs black and deep; now it opens out into a broader and shallower bed dotted with banks and islands. Its vegetation is distinct from that of the country from which we came".

Livingstone was not alone in his enthusiasm. Anderson wrote of the route to Kanye in 1869, probably from Pioneer Gate, the road "passes through a more lovely country than the one just described, the rich alluvial soil of the valleys, well watered by fine springs, which are small branches of the Notwane river, finer grass-lands studded with beautiful groups of trees and bush. On every side of the road, well-wooded, lofty, and picturesque hills-they may be termed mountains; others in the distance rearing their lofty heads, visible between the openings of those near. The subtropical plants, scarlet creepers climbing up and between the isolated rocks, piled one upon another, complete a landscape seldom to be surpassed for the beauty of its scenery."

Between Kanye and Masopa the country was "very picturesque, lovely valleys, some well cultivated; many of the hills that surround them are clothed in lovely vegetation-the euphorbia, wild fig and other subtropical plants; creepers of every variety climbing up between the large masses of sandstone rocks that stand out in grotesque forms, piled one upon another, add much to the beauty of the landscape. Such charming scenery should not be passed over in haste".

Even the pessimistic Bryden wrote: "And yet during the brief weeks of rainfall no land can assume a fairer or more tempting aspect. The long grasses shoot up green, succulent, and elbow-deep; flowers spangle the veldt in every direction; the giraffe acacia forests, robed in a fresh dark green, remind one of nothing so much as an English deer-park; the bushes blossom and flourish; the air is full of fragrance, and pans of water lie upon every hand.

"Another month and all is drought; the pans are dry again, the grasses are turning to their winter yellow".

Some could get excited about much smaller delights than big game. Dr. Emil Holub, a Czech, was an avid naturalist who made several trips through eastern Botswana between 1873 and 1877. He collected 18,000 insects, nearly 300 reptiles, 782 mollusc shells, 400 birds, 3328 plants; fossils, minerals and other items; not all in Botswana but some in South Africa as well. Of the country between the Shashe and the Tati rivers he wrote: "finding the scenery where we quitted the river [Shashi] as beautiful as any in the whole West Matabele country. The abundance and variety of plants were truly marvellous; and on the slopes where the stems of the euphorbias were mouldering, I found numerous scolopendra, two kinds of scorpions, some lizards, and many sorts of insects."

Thirty years after Cumming and Methuen, at the beginning of 1876, Parker Gillmore hunted over much of the same ground. In Livingstone's day Mabotsa had been infested with lions but now "You will not hear a lion until you have crossed the Notawaney River", Gillmore quotes, although pointing out that they were still present in the Transvaal. The Notwane did not flow in the dry season, but had a succession of deep, stagnant pools, some a couple of hundred yards in length. After crossing the Notwane all day large antelopes were in sight, but Gillmore remarks that elephants and buffaloes had become scarce since Cumming's day.

Of the Limpopo between Buffel's Drift and Parr's Halt he wrote: "its clear waters, wooded banks, and constantly repeated pools and rapids-here and there interrupted by beds of rocks or massive boulders-give it a bewitching air of enchantment.

"The first time I viewed this portion of the charming river, on the opposite bank basked several crocodiles, while francolins and guinea fowls dusted themselves in the loose sand. In the adjoining trees was a quantity of the pretty little grey monkey, familiar to the neighbourhood, while a little farther down its course drank about a dozen of the awkward-looking, but powerfully-built bastard hartebeest [tsessebe]. Elephants had been here lately; their spoor was to be seen all over the neighbourhood, and broken and felled trees attested to the strength of this mighty rover of the forest. "The hippopotamus is also found, but not in such abundance as in years gone by".

2.3 A Most Abominable Spot

In the middle of the southern Africa winter, when the northern hemisphere is enjoying its summer, an incessant cold wind blows for days on end, covering the land with a thick, grey fog of dust in the region of the great Makgadikgadi salt pans. At times you can barely see for more than a few yards before you in this bleak, grey world of dust and drought. In 1988 the Boteti River, reached after a long, waterless traverse of the salt flats from Nata, was but a pitiful remnant of the river Livingstone described in 1849 as "beautiful", the river which was to lead to the fabled Lake Ngami after crossing the near waterless desert from Serowe. Selous did not think much of it in 1879: "we at last reached the Botletlie or Lake river… This is one of the most abominable spots I have yet visited; one small mud-hole from which a little filthy water was all we could get for ourselves or the oxen, and yet on the map this river looks like a young Mississippi. At certain seasons it is doubtless full, but just now [May 4th] it is a horrible parody upon a river… we crossed the river at a very good drift, the river being quite dry". In fact he was very high up, at Mopipi near the termination of the water's seasonal flow.

The adventurous Sub-inspector of the Bechuanaland Protectorate Police, Arnold Hodson, visiting it much later still in May 1906 was also not impressed: "we arrived at the Botletli, full of eager anticipation. But alas! no flowing Thames or majestic Mississippi broke upon our gaze… One could see no drop of water anywhere, nothing but the dismal expanse of dried and withered reeds… Just at this point water is only to be obtained from a few muddy holes during the dry season".

When Bryden visited it in 1890 the vast reed beds of Lake Xau were being burnt: "The reed-beds at this end of the river stretch for miles over a kind of swamp, in which the Botletli loses itself [Lake Xau], and as each reed stands some twelve or fifteen feet in height, and is highly inflammable, some idea of the sight of acres upon acres in conflagration may be imagined.

"At this time of the year [July] the Botletli, which rises so mysteriously in the middle of the dry season, was low at this end, and showed but a comparatively small channel of water. As we trekked... up the river, however, we found the water rising, and a great stream, overflowing its banks, flooding the country round, and forming vast lakes, channels, and lagoons in every direction."

When he saw it, Lake Xau was covered with flamingos and pelicans, the latter especially numerous. He was puzzled by the flooding of the Boteti. On the 23rd of June 1890 he wrote: "The river is very full, and still rising rapidly, overspreading its banks far and wide… And yet not a drop of rain has fallen for months… I cannot in any way account for it-it is one of Nature's freaks."

Bryden asserted that the yearly inundation was not losing its volume as some supposed (the white community at Maun thought the same in the 1980s drought), but was as copious as ever "and the mass of water has suffered no diminution for fifty years past. In some seasons, indeed, it would appear to have exceeded its modern limits. Beginning in the month of April, the inundation slowly proceeds until, late in the month of August or early in September, the highest point is attained, the lower reaches are flooded, and the subsidence begins… Lower down, near Lake Komadau [Lake Xau], the channel is at this time narrow and shrunken; yet there is always water-occasionally deep pools-between the flat, dry mud-banks even here."

Crocodiles were numerous. Somewhere in the Mopipi region Chapman in 1854 recorded seeing as many as 57 huddled together on the margin of a pool where women collected water, and one that he killed was some 4.9 m in length. Eight men could not drag it out of the water. On another visit he was "truly horrified" at the sight before him, for this time he saw not 50 but 200 crammed together in a small pool, the river at this season having shrunk into pools. One he shot was 5.2 m in length. The San had no scruples about eating the flesh, but at Lake Ngami, where crocodiles were numerous, the people held them in superstitious awe and believed if they were killed the river Thaoge feeding the lake would dry up. As it subsequently did.

Large crocodiles still exist in the Boteti, able to pass the dry season buried in the mud. A few muddy stretches were all that remained at my visit and the east bank was bare, stripped of its vegetation by innumerable hungry cattle. It did not look any different to the bare ground in a photograph taken by Chapman in May 1862 (see Tabler, E.C. 1971) when I photographed it from the same spot in July 1987. When the herds of zebras, wildebeests and buffaloes came there to drink in 1988 there was no food left for them by the cattle and they died. Died in their hundreds. I passed a group of three young four-year-old buffalo resting lying down in the sparse cover of the thorn trees which afforded no shade to the exhausted beasts. As I approached they scrabbled to their feet in fright and galloped away. At least two of them did, the other stumbled and collapsed after a few steps to lie kicking and rolling in the dust. Further on I came to another, lying miserably with eyes closed, it's ear all bloodied, half torn off by a hyaena as it lay, too weak to try and rise when I approached (Plate 2.2).

Plate 2.2 The Boteti river, July 1987

2.4 Hell upon Earth

The early travellers did not record such sorry sights, but while frequently enthusing about the quantity of game were often unimpressed by the countryside. Even Livingstone, accused of painting an excessively benign scenic picture, while cutting cross-country from Nxai Pan to the region of Kudumane in 1850 wrote: "It is impossible to convey an idea of the dreary scene on which we entered after leaving this spot [some pools to the north-west of the Nxai Pan Park]: the only vegetation was a low scrub in deep sand; not a bird or insect enlivened the landscape. It was without exception the most uninviting prospect I ever beheld". And of the country north of Mahalapye he was to write: "The pass was also called Porapora (or gurgling of water), from a stream having run through it. The scene must have been very different in former times from what it is now [1852]. This is part of the river Mahalapi, which so-called river scarcely merits the name… The country beyond consisted of large patches of trap-covered tufa, having little soil or vegetation except tufts of grass and wait-a-bit thorns, in the midst of extensive sandy grass-covered plains. These yellow-coloured grassy plains, with moretloa and mahatla bushes, form quite a characteristic feature of the country. The yellow or dun-colour prevails during a great part of the year… The grass grows in tufts the size of one's hat, with bare soft sand between. Nowhere here have we an approach to English lawns, or the pleasing appearance of English greensward."

William Baldwin, who had been captivated by the books of Harris and Cumming and ranks with them as one of the great nimrods of the day, expressed his disappointment in no uncertain terms when travelling westwards south of the Makgadikgadi

Pans to Lake Xau in 1858: "I am much disappointed both in the country and the game. There is no variety whatever here, only eland, giraffe, and blue wildebeest with a sprinkling of duikers, steinbuck and springbuck, the kinds that are least dependent on water, and even these are few and far between."

His impression of Nata in October 1860 was no better: "a more comfortless, barren, desolate spot no human being could conceive. There is neither grass, wood, nor water; the sun is intensely hot, and there is no shade of any sort".

I know how he felt, for I once broke down on the road across the open salt flats west of Nata and was stuck there for 24 h with not a tree in sight to offer shade during the heat of the day.

But Baldwin was rather like those modern tourists who complain of having seen "nothing" in a national park if they have not seen lions. In July 1859, hunting in the Nxai Pan region he came across elephants watering at a pan: "Thus we bagged eight in about half an hour. A few clumps of giraffe, a troop of tsessebe or buffalo, or a white rhinoceros occasionally crossing our path, with a small lot of gemsbok and a few quaggas [zebras], we did not deem worthy of notice." He went on to complain: "The country is almost devoid of game of any value, zebras and blue wildebeests being the only varieties. I shot an old bull giraffe three days ago, about the last of his race in these parts."

The giraffe fortunately still occurs there today.

In August the following year the missionary, John Mackenzie, saw things rather differently. A little east of where Baldwin shot "the last giraffe" in the region near Gweta he wrote: "The game was here more plentiful than we had before seen it. Vast herds of buffaloes rushed past our waggon as we approached the water [a fountain called Maila]. Troops of zebras and gnus were to be seen in several directions. The smaller antelopes abounded; the tracks of rhinoceros were numerous; while elands and giraffes grazed not more than two miles from our waggons."

In 1875, at the beginning of the rains when Barber camped near Makwa, north of Serowe, the flats were "covered with game, assembled from all sides to graze on the green young grass." Meeting a party of Khama's hunters they asked him to shoot as much as he could for them as the harvest had been bad. Barber records that he came upon a great herd of zebra "I am afraid to say how many, there were hundreds".

Schulz and Hammar on the eastern shore of Sowa Pan in 1884 reported that "Here many kinds of game congregate in numbers: elands, blue wildebeest, quagga [zebra], and springboks in large herds are the principal sorts we met with."

Robert Moffat's son John did not visit Botswana until 1863: "We are now at the Lobatse River, another lovely spot. These hills and dales once resounded with the voices of tens of thousands of inhabitants, when the Bechuana nation was still in prosperity. The ruins of stone kraals and other enclosures spread around us for miles, now choked with rank grass and thorny copsewood and still as the grave. Water must have been much more abundant in this region then, for this enormous mass of inhabitants, with their cattle, could not have subsisted here in the present state of the country. What a curse seems to rest on this land!... There is something very dreary about these people-no sentiment, no love of nature, no tradition".

on their way to Lake Ngami; the former two hunting and the latter trading. The first Afrikaners, Jan Viljoen and a party of others, reached Lake Ngami in 1851. In September 1852 Chapman met a trader called Simpson near Letlhakane returning from the Linyanti, all of whose trek oxen had died of *nagana* (trypanosomiasis, carried by the bite of an infected tsetse fly *Glossina* sp.), and when he reached Shoshong there was another Englishman outspanned there named Mr. Reeder. In June 1854 he met at Nkokwane, south of Ntwete pan, Messrs. Dyme and Moyle, "travellers who had left the Cape two years previously, and had encountered great disasters in attempting to reach the lake by a road parallel to that followed by McCabe through the Bakalihari desert, losing all their cattle and horses from thirst". Joseph McCabe had attempted to reach Lake Ngami with Baines in 1850, and in 1852 was the first to cross the central Kalahari with ox-waggons, taking the route north-west from about Molepolole to Lake Ngami. There was an abundance of tsama melons that year and his oxen subsisted on them for 21 days without needing water to drink. Shelley, accompanied by Webber and Richard Orpen, tried to follow, taking a route further west to Dutlwe and Kang. After diverting westwards to Lehututu for water they were forced to abandon the attempt. An English traveller hunting and trading, Moyle was allegedly the first white man to reach Ghanzi, crossing the desert in 1852 and afterwards making his way to the west coast. But according to the date of 1851 carved on the Gutsha baobab, Hendrik Van Zyl may have been the first. A party of Griquas had reached it many years previously, having abandoned their waggons in the desert. During 1855–56 the trader Frederick Green accompanied Dr. Johan Wahlberg, a Swedish naturalist collecting for the Stockholm Museum. They went to Lake Ngami and up the western side of the Okavango Delta to Seronga, then eastwards until Dr. Wahlberg was killed by an elephant; perhaps in the Tamafupa area. On October the 5th 1859, somewhere between Bushman Pits and Gweta, Baldwin met a Mr. and Mrs. Thompson on honeymoon, accompanied by a Mr. Palgrave. They had come up from the Cape and were crossing to the West coast (Walvis Bay). Also present was a Dr. Holden with two white servants, who intended going to the Zambezi and then south through Zimbabwe. About a week later at Letlhakane he met the English missionaries proceeding to Linyanti, Holloway Helmore and his family, and Roger and Mrs. Price. All, except Roger Price and two of the Helmore children, were to die at Linyanti, probably of malaria. In July 1860 at Mpandamatenga he met another Englishman, Polson, who had crossed from Walvis Bay buying ivory. At this time he learnt that Dr. Holden was now living somewhere on the Chobe River. When he reached Serowe, "greatly to my astonishment", he met two Englishmen, Waddington and Aldersley, who had also crossed from the West coast just for the adventure. He also heard of three Englishmen hunting, one of whom was blown up when all of their gunpowder, 720 kg of it, exploded. At Nata in 1860 he was shown the grave of an unknown white man. In 1870 an Englishman named Hart married in Durban and took his wife on a honeymoon to Tati, combining his honeymoon with a hunting expedition. Eduard Mohr commented "none but a native of Britain could have conceived it!"

By 1874 tourists, as well as hunters and traders, were already visiting the Victoria Falls, reaching there by ox-waggon from Pretoria then via Shoshong, Tati and Pandamatenga. It was estimated that 12 would visit the Falls that year. In 1858 Wilson, the trader who had accompanied Livingstone on his first journey to Lake Ngami, was living there. Later, about 1866, Strombom, a British subject of Norwegian extraction, established a store at Lake Ngami for a period. He died there on a visit about 1892. By 1877 Harry Shelton had also established a large store, but it was still just as hazardous a journey in the dry season to reach Lake Ngami as it had been in Livingstone's day.

3.2 The Animals' Pains

"I am sorry now for all the fine old beasts I have killed", wrote Oswell in 1893. But in 1914 a leading publisher of boys' books brought out Sir Harry Johnson's Pioneers of South Africa with a coloured dust-jacket depicting a man galloping into herds of blesbok and zebra, shooting from the saddle. Good fun for boys! (Plate 3.1).

Cumming has been criticized for his seemingly insatiable bloodlust and indifference to the sufferings of his quarry. Yet he was a product of his time and the measure of a sportsman in Britain was how many animals he could kill. This was not to everyone's liking, and Cumming's shooting hippopotamuses on the Limpopo between Martin's Drift and the Lotsane confluent, where he was "struck with astonishment" at their numbers, led to one reviewer of his book writing: "But the slaughter

Plate 3.1 Good fun for boys-a 1914 view

of the poor helpless brute is too disgusting a story for our readers, and is one of the numerous instances of bad taste afforded by the book."

In *Punch* it earned him satirical criticism under the title "The Gordon is Cumming, A Roar from a Wild Beast":

> "The GORDON is CUMMING, oh dear, oh dear!
> The GORDON is CUMMING, oh dear, oh dear!
> To slaughter us wholly; that's clear, quite clear.-
> (I've a bullet of his in the rear, the rear).
> He'll cut us all up, branch and root, and root,
> No creature he spares from pursuit, pursuit:
> The King of the Beasts he will shoot, will shoot,
> And the antelope also, poor brute, poor brute!
> He blows out the elephant's brains, his brains,
> His hand with giraffes' blood he stains he stains,
> As he in his volume explains, explains,
> Disregarding the animal's pains, its pains.
> The sea-cow he peppers, pop, pop, smash, smash!
> She flounders and rolls in her gore, splash, splash!
> At last goes a ball through her skull, crash, crash!
> What a mercy it settles her hash, her hash!
> Perhaps we were made with intent, intent,
> That balls through our sides should be sent, be sent,
> Our nerves were contrived to be rent, be rent,
> And our bones to be shattered were meant, were meant.
> If that is why we were sent here, sent here,
> Of course 'tis all right-though 'tis rather queer-
> And to put us to use in our sphere, our sphere,
> Mr. GORDON is coming-oh dear, oh dear!"

Others were no better. Henry Methuen with three companions; Bain, Pearson and Monypenny, hunted over much the same area at the same time from late 1844 to 1845, entering Botswana on August 19th 1844, Cumming having entered on the 4th of June. Methuen's book, Life in the Wilderness, was not as exciting as Cumming's and is less well known, but his bloodlust was just as bad: "I wounded and lost several rhinoceroses, the bush being very thick", he casually remarks. Also, having fired into a herd of buffalo and wounded two "according to a maxim well known to sportsmen here, 'never follow a wounded buffalo', we left them alone".

A little over ten years later William Baldwin, as bold a hunter as Cumming, slaughtered just as much when he could: "Three giraffes, three white rhinoceros, one black ditto, and one eland bull, must be added to the list of slaughter" he wrote in October 1857, north of Palapye. South of Ntwetwe salt pan (1858): "I saw a small troop of five elands from the waggon, and, I grieve to say, we killed them all; but as some excuse for our wholesale butchery… there were about 150 Kaffirs to provide for, and not a particle was wasted."

On the Notwane river Chapman wrote in his diary in May 1852: "We saw abundance of game and slew to our hearts' content."

3.3 Only Seven Lions Killed

Cumming's encounters with lions were numerous, they appeared to abound everywhere. But Chapman, who also made frequent reference to them, wrote: "This makes only seven lions killed by me during twelve years of experience of African travel, so much less numerous are these animals than is supposed by many persons." Baldwin listed only four killed on his last expedition, and Selous was to write in 1874 that on the bank of the Tati river he at last shot his first lion: "Many people may think it strange that I should have been so long-nearly three years-living almost entirely in the wilderness, in countries where lions were plentiful, and where I was continually hearing them at nights, and yet never have seen one; but such is the case… Being nocturnal animals, and accustomed to lie asleep in beds of reeds and thick patches of bush during the daytime, it is mere chance work ever coming across them".

On his hunting trip to Matima on the Boteti and back in the dry season of 1890, Bryden saw the spoor of lions around his camps a few times on the Boteti, but never saw a lion itself during his whole safari.

In 1861 Chapman wrote that the San, using poisoned arrows, had killed off all the lions between about Gobabis in Namibia and Lake Ngami: "and not a lion is to be seen in the whole country between Amraal's and the Lake". But the lion does not disappear that easily.

Holub considered there to be three species of lions. The full-maned lion, the maneless lion and a kind with a mane which never reached below the shoulder. The "spotted lion" he was aware lost its spots with age. Full-maned lions were rarely seen, it generally being assumed the hairs were pulled out by thorn bushes. Maneless lions in 1875 he asserted used to be common on the Molopo river, and were still found on the lower Chobe, their colour being extremely light.

The most common species was the short-maned. Today the full-maned lion is considered characteristic of the Kalahari. Selous did not believe more than one species existed as the coats of the male were very variable in colour, size and length of mane, and there was a gradation in size of mane from maneless to big-maned ones. In 1879 he shot two "fine old male" lions on the Mababe flats, lying together under the same bush. One was full-maned with a very dark-coloured coat; the other very light-coloured with hardly any mane at all. Both were approximately the same size. Neither did he agree with the view that the hairs of the mane got pulled out by thorn bushes, for you could find maneless ones where there were no thorn bushes and lions with thick manes where there were thick thorny jungles, as near Tati.

Leyland had written in 1851 that it was the old lions were the most dreaded by the people and caused the greatest destruction to human life. No longer able to catch or overpower the larger animals they had no hesitation in attacking man or domestic animals. Livingstone subscribed to this view, noting that when a lion got too old to catch game it frequently took to killing goats in the villages and a woman or child happening to go outside at night could fall prey to it. As this was now its only source of sustenance it continued, rather than once having tasted human flesh it preferred it to any other. He stated that a man-eater was invariably an old lion and the natives had a saying "His teeth are worn, he will soon kill men." Away from human habitation such old lions begin to catch mice or other small rodents, and even eat grass. The San, seeing vegetable matter in the droppings, would follow the lion up and kill him without difficulty. Others considered lions to be "especially dangerous" between the ages of two and four years. Holub noted they were much less dangerous in places where their food was plentiful. Where they were much hunted and had become familiar to the sound of fire-arms, as well as in areas where game was scarce, then they were much more dangerous, attacking humans and livestock.

Livingstone pointed out the peculiar fact that the roar of the ostrich was very like that of the lion and he never knew a European who could distinguish the two apart. One could only guess which was making the sound. If during the day it was probably an ostrich, if at night, then a lion. The locals asserted there was a slight difference in the commencement of the roar enabled one to tell the difference, but none in loudness or tone. To Selous however the two roars were as different as the notes emitted by a concertina and a cathedral organ. The booming noise made by the cock ostrich during the breeding season sounded, at 45 m, very like the roar of a lion heard at 5 km.

3.4 The Tables Turned

Mackenzie found hyaenas to be very bold at Shoshong in 1862, and in 1824 Moffat had recorded that hyaenas around Kanye commonly attacked and killed people, often seizing children who were drawing water in the evening. This may have resulted from smallpox epidemics, when hyaenas became accustomed to feeding on the bodies of the dead left in the bush. Smallpox was first introduced to the Cape in 1713 and several outbreaks followed in the eighteenth and nineteenth centuries, with epidemics in 1812, 1831–2 and 1858.

This practice of leaving corpses in the bush had not always been the case. Henry Lichtenstein, writing of the country at the beginning of the nineteenth century, although he did not cross the Molopo, noted the Tswana did not, like some tribes, abstain from eating hyaenas because they ate corpses, for amongst the Tswana they did not. The dead were buried in cattle kraals and not, at that time, left out in the bush.

3.5 The Fall of the Giraffe

On the west side of the hills near Otse, Methuen saw giraffes, elands, kudus, impalas and buffaloes, as well as a number of rhinoceroses. After the shooting of a bull and a cow giraffe by himself and one of his companions, he noted: "Sportsmen here seem to be great murderers" but so many half-starved natives besought us to shoot for them, and such was the waste and epicureanism amongst our own Hottentots, that we could hardly kill too much game."

The pans at Boatlaname and at Lephephe were favourite localities for the giraffe in the 1860s, and Anderson shot two out of seven some 13 km from Boatlaname. The trader Hume on another occasion shot three from the saddle: "It is a pity such beautiful animals should be destroyed merely for food and skins" he wrote.

Even the great Selous described shooting a giraffe in 1876 as follows: "She was not yet dead, however, for as I approached she reared her lofty head once more and gazed reproachfully at me with her large soft dark eyes. For the instant, I wished the shots unfired that had laid low this beautiful and inoffensive creature." His companion had killed another nearby: "As the one I had shot was the fatter of the two, we left the Kafirs here and went back to mine."

On the next day they sighted two more giraffes and "had a glorious run after them" killing a young bull: "being now well supplied with meat, we resolved to shoot no more giraffe until it was all finished." But the next day at the request of some San who came to the waggons saying they were hungry, they went out again and two more cows were killed. Then the next morning some Ndebele warriors appeared and requested meat, so a solitary old bull was shot; and the following day another three cows. Not much remorse there! But in 1905, at the age of 57, he was to write to the Reverend J. Stratton who was campaigning against the Eton Beagles, "After reading your pamphlet I certainly think it would be better to substitute drag-hunting for the pursuit and killing of a hare. To see one of these animals worried and torn by a pack of dogs is not an edifying sight for a young boy."

In 1852 Chapman recorded a herd of 200–300 giraffes "migrating", moving out of the desert somewhere south-east of Makoba. Near Mahalapye in 1865 Finaughty, together with Khama and the Reverend Roger Price, coursed a herd of 300–400, killing five. At the beginning of April they moved out onto the plains to eat the sweetish but stringent *mogwana* berries (*Grewia bicolor*).

Chapman both hunted, collected natural history specimens, and traded, over a period of ten years from 1852 to 1863. Giraffes were one of the commonest species shot by him and all of the other hunters, being a favoured meat and usually carrying a goodly quantity of fat. East of Shinamba Hills in 1863 he came across a herd of about 20 which he remarked were "as white as ghosts", apparently a sure sign that they were fat. This is not an observation that is ever made today. In the region of the Botete River he was informed by a local chief white buffalo were sometimes captured in their traps, and white zebras without any indication of stripes frequently so. These animals were not white for the same reason as the giraffe, but albinos also are rarely reported today, presumably because of their normally low frequency in any population.

In 1879 Major Serpa Pinto wrote that giraffes abounded in the Kalahari and he occasionally saw herds of them, "but never once succeeded in killing".

Selous found in 1881 the range of the giraffe had been sadly reduced since Cumming's time, but nevertheless the giraffe was still to be found in considerable numbers. It was said to abound in parts of the Kalahari desert and was very numerous between Shoshong and Lake Ngami, and up to the Chobe river. It was particularly plentiful along parts of the Boteti and in the forest-clad sand-belts on the southern bank of the Chobe.

In 1890 Bryden considered the best giraffe country was the northern Kalahari abutting on the Boteti river and on the Mababe flats to the north. But he first encountered them near Maakwa. South of the Makgadikgadi Pans he came across a herd of 19, "It is seldom nowadays, even far up in the hunting veldt, that one happens upon so large a troop of "camel"… Right in the Kalahari, some days south from the Botletli… large troops yet wander… I have heard of seventy and eighty giraffe being seen at one time in that country even within the last few years; but such sights are, I imagine, of rare occurrence."

"But year by year it has been hunted from one sanctuary to another, until now the giraffe must be sought in the far corners of Khama's country, or in the waterless recesses of the North Kalahari, before it can be found in any numbers."

In 1889, on the way to Lake Ngami Nicolls's party shot three near Matima Ford on the Boteti, and of the return journey near Makalamabedi the following year, in an article in The Field, Nicolls recorded: "This wound up the most exciting day's sport I ever had in my life. Strombom had shot two bulls and a cow; Hicks one bull and a cow; and I myself a cow, which I had left for dead on the veldt, but, as the bushmen who had been running on our spoor informed me, got up again and ran fully half a mile before they could come up and polish her off with their assegais." By way of explanation he added: "Although we killed a large quantity of game, still I can say this much for my friends and self, that on no single occasion did we shoot a head of game for the mere purpose of wanton slaughter."

In 1892 the same author wrote: "The prodigies of slaughter may be conceived when it is stated on reliable authority that nearly three hundred Giraffes have during the past two years been exterminated in the'Ngami country merely for the sake of their hides."

Probably on account of drought in 1890 "a troop or two" moved south almost to the Molopo River. It is doubtful any ever reached it because local hunters from the south flocked to meet them. Only on the south bank of the Chobe River and near the Okavango River did they remain numerous. Khama controlled a very large portion of the best giraffe country left in southern Africa at this time and Bryden thought he might be induced to protect "these rare and harmless animals". Already Khama only allowed his own people or his friends to shoot them, and five years later he issued a decree prohibiting hunting giraffes without his permission. The same year, together with the chiefs Bathoen and Sebele, Khama presented a giraffe skin to Queen Victoria when she met them during a visit they made to Britain (Plates 3.2 and 3.3).

Plate 3.2 In 1890 Bryden believed "this rare and harmless animal", the giraffe, would soon become extinct

Plate 3.3 "The fall of the giraffe". The Illustrated London News, March 8, 1884

3.6 The Best Wwaggon-Grease

South of the "Mountain of Eagles", probably Mannyelanong Hill near Otse which Bryden in 1890 referred to as "a headland celebrated from time immemorial as the resting-place of vultures, and noticeable at great distances in the surrounding country", Cumming shot his first eland: "Like the gemsbok, the eland is independent of water, and frequents the borders of the great Kalahari desert in herds varying from ten to a hundred. It is also generally diffused throughout all the wooded districts of the interior where I have hunted. Like other varieties of deer and antelope, the old males may be found consorting together apart from the females, and a troop of these, when in full condition, may be likened to a herd of stall-fed oxen. The eland has less speed than any other variety of antelope" (Plate 3.4).

The eland found in the south of Botswana *Taurotragus oryx oryx* lacks any white stripes, while northwards is found an intermediate form between the southern and the striped Livingstone's eland *T.o. livingstonei* of Zambia. In the intermediate form the young are born with nine stripes, more distinct on the forepart of the body, but these may or may not disappear either in part or in whole as the animal grows older, producing a very variable pattern, even in the same herd. The eland which Baldwin shot at Tamasanka Pan in 1860, and which he claimed was a Livingstone's eland, was probably an intermediate form. Nicolls placed the dividing line at the Boteti River, claiming those on the south bank were a uniformly bluish-grey dun colour, while those frequenting the northern bank were marked more or less indistinctly.

Methuen, hunting in 1846 on the South African bank of the Molopo, described how to hunt the eland: "In the Kgalagare the elands, which require little or no water

Plate 3.4 "Mountain of eagles", nesting site of the rare cape vulture at Mannyelanong, an early landmark on the Botswana border

where the water-melon grows, still exist in vast herds. The best method of shooting them is to ride in pursuit, keeping behind and to the left of them, with your horse well in hand, lest he should turn suddenly upon you, and when within a yard of their tails, discharge your gun into their flanks, wheeling off to the left at the same moment."

"Upon the fat, which is found in large quantities on the breast and by the heart of the eland, we entirely depended for waggon-grease, candles, and many culinary delicacies."

Livingstone told him elands were reported to abound in the desert; "and the soft, yielding sand, with the extreme scarcity of water, proves to them, and to the other animals inhabiting those parts, a sure defence against the murderous expeditions of the hunter."

On another occasion Methuen described the Notwane as "flowing in a deep bed, the water hidden from view by high flags and jungle" where he saw the spoor of rhinoceroses, buffaloes and other game, and of one elephant. In the hills on the other side he shot a klipspringer. Four bull elands were seen, of which two were shot as their fat was needed for "waggon-grease, candles, culinary purposes, and rendering our leakey tent water-proof." Five days later "a few elands were observed", and he notes "these valuable creatures not having been as frequently met with as we could have wished."

Livingstone observed: "It is remarkable that attempts have not been made to any extent to domesticate some of the noble and useful creatures of Africa in England. The eland, which is the most magnificent of all antelopes, would grace the parks of our nobility more than deer. This animal, from the excellence of its flesh, would be appropriate to our own country". Farini, in an address to the Royal Geographical Society in 1886 (read in his absence) stated: "The vast herds of elands roaming the plains might be domesticated and herded like cattle, producing as much meat, with less cost and risk."

Many trials of eland domestication have been made from Victorian into present times, but to put on the weight that they do they require a rich diet and trails have been found not to be economically worthwhile.

3.7 Game by No Means Abundant

Along the Marico Methuen met his first waterbucks, a herd of does, later seeing a number of bucks. The horns of one he shot measured 74 cm. The locals did not refuse the flesh although he noted it was currently said the flesh was so rank as to be rejected even by them; a story met with today. The buck did not have a strong smell, but he thought this might depend upon the season.

Crocodiles and hippopotamuses were numerous in the Marico, and of the hippopotamus he wrote "Their feeding time is at night, when they crop the grass and small shrubs along the river margin as closely as if they had been mown'. Interestingly enough he also records that "at this time they also migrate, when from any cause their residence has become untenable, and pass over long distances with extraordinary celerity... they are said, on these occasions, to steer in a direct line for the nearest water with an unerring instinct."

Nevertheless he remarked "Game is by no means as abundant here as we had expected, and by the old tracks seems to have departed for some other quarters about a month back" [i.e. August, rather it had now dispersed away from the water because of the first rain showers]. But "The emigrant boors [*sic*] make constant shooting excursions up the Mariqua, for elephant and hippopotamus ivory, which accounts for the latter animal being so scarce and wild."

Meeting a party of Afrikaners hunting on the Marico and who, shooting a rhinoceros, took only the hide and a small amount of fat and meat from the ribs, informing him they would shoot three more if they could just for this small amount of meat, for, next to the eland, it was their favourite food; he pointed out to them the wastefulness of this, dependent as they were upon game. They replied the supply was inexhaustible. Just as almost 50% of Batswana stated in modern times, believing, as did the Afrikaners, the animals were, and would be, put there by God. After his homily on the subject of wastefulness, Methuen promptly killed a rhinoceros himself to try out a shot the Afrikaners had shown him, and was obliged to also shoot the calf because it defended its mother.

Baldwin hunted giraffes probably in the area west of Kgale Hill, and also shot two impala there. Travelling from Kopong in September he noted there was no water for three days until Boatlaname, and complained of the scarcity of game, although he hunted some giraffes in the area. Reaching Lephephe he could find no animal spoor, but nearby found three black rhinoceroses of which a cow and her calf were killed. He saw sable antelope near the Shoshong Hills, but it was probably near the Tati River that he shot his first, and his first tsessebe, also finding roan antelopes there. Of the Tati district Baines wrote in 1869: "Game of all kinds, including the various antelopes, buffalo, eland, giraffe, lion, rhinoceros, and elephant are to be found in the Tati district, but the professional hunters make them, especially the elephant, scarce and shy"

3.8 The Whole Sordid Business

Cumming is particularly vilified for practising shots on a wounded elephant after he had rested and brewed a cup of tea by the immobilized, suffering beast. But Holub in 1875 referred to a South African hunter named "Henry W.": "I could not get over the barbarities which he permitted during his excursions." On one occasion, after wounding a cow elephant which had chased him, he allowed his servants to torture it for a couple of hours with spears before he shot it.

But just as bad, if not worse considering our alleged age of enlightenment, happens today. In 1984 an infamous case in Botswana concerned a hunting safari from South Africa notable for the presence of a renowned surgeon as a member. He was to provide the commentary for a privately sponsored film to be made by a wealthy South African sheep farmer to publicise hunting. Twelve white "hunters", consisting of nine from South Africa and three Botswana farmers, (of whom only two of the South Africans had licences, one being the surgeon), went in search of two lionesses with three cubs which had been spotted from the air.

On the way the convoy of vehicles encountered a honey badger, a protected species under Botswana law, which after filming they set their dogs onto. The honey badger however gave a good account of itself before escaping. Continuing on its way the convoy came across a brown hyaena, another protected species and internationally endangered. The hyaena was chased by the vehicles to be lassoed for filming. In the pursuit the hunters bumped it several times "accidentally" and eventually, as it was being lassoed, it was run over by one of the vehicles and died shortly afterwards of internal injuries. The sexual organs were displayed and photographed and the discussion so beloved of the ignorant took place with reference to the hyaena's apparent (but false) hermaphroditism, before the corpse was left in the bush.

At last the unfortunate lionesses were encountered. But as they showed little interest in their visitors, attempts were made to provoke them into charging by chasing them with the vehicles. Not wishing to play, one lioness suddenly stopped and crouched down. The vehicle ran into her jamming her under the front bumper and becoming stuck. One of the other drivers then pushed the vehicle over the lioness's body with his vehicle, and for good measure ran over the lioness himself. The injured animal was then shot dead by the surgeon from within the safety of the vehicle.

Meanwhile the other lioness had apparently been deliberately wounded with several shots fired into her at close range to try and provoke a charge. Drink cans were thrown at her and she made a weak attempt at leaping at the front of the vehicle when the surgeon shot her, again heroically from the safety of the vehicle. But the fun was not over yet, and the party then chased the three now orphaned cubs, estimated at about 15 months old, and lassoed one, dragging it to the vehicle bumper. Here one brave hunter poured beer over its head and nudged it with his foot. The cub was then released but there was little chance of any of them surviving.

Armed amazingly with a portable sausage-making machine the safari continued: killing gemsbok, springbok, wildebeest, hartebeest, duiker, eland, tsessebe, reedbuck, lechwe, impala, warthog, buffalo and zebra. Turning them into sausages as well as biltong. Many animals were wounded and left, and on one occasion two of the South Africans, neither of whom had licences, allegedly fired into a herd of some 400 buffalo, wounding four.

Attitudes haven't changed as far as some people are concerned, but even the much maligned Cumming would not have been so gratuitously cruel as these people proved themselves to be. In the 1980s some farmers of South African origin in the Ghanzi area competed with one another to claim who had shot the most lions. Most of the killings were illegal, some taking place in the Central Kgalagadi Game Reserve; but there was a ready market for the skins in Namibia.

As to quantity, in 1972 a Botswana Member of Parliament and honorary game warden, accompanied by two district councillors, all without licences, in two days shot 52 zebra, 17 wildebeest and five gemsbok. This was in the Makgadikgadi Pans Game Reserve where hunting was forbidden.

None of the early sportsmen who have been so criticized shot such numbers in a matter of two days, although the Afrikaners undoubtedly did so of animals migrating in the former days of plenty. But Botswana has suffered from a legacy of senseless slaughter. In 1953 the administrative officer based at Francistown in charge of the Northern Protectorate, stated "the whole sordid business is out of hand". He was referring to hunting by the Batswana, or "citizen hunting" as it is termed. Having stopped the issue of export permits from Francistown, only one of the ports of exit, in one week he had held up 3.5 tons of hippopotamus skin strips; 173 buffalo hides; 227 other game skins, and a further ten tons from Nata equal to 2000 skins. In that year over 6800 springbok skins were exported from the Southern Protectorate. Lacking the conscience of her forebears, Mrs. Moremi, the Batawana Regent, issued in 1952, 44 elephant permits, 293 hippopotamus permits, 144 giraffe permits, and 151 eland permits. All specially restricted species in Ngamiland since 1910.

Members of a Zimbabwe museums expedition to Makgadikgadi in 1953 reported at Francistown the "terrible, and in their opinion, disgraceful slaughter of game in areas they have crossed... They state the plains are strewn with wounded or dead wildebeeste and other game which has been "browned" and wounded at water holes. They mentioned a koodoo bull which they came upon freshly killed with the meat untouched and only the hide removed. Native hunting parties are everywhere".

It seems nothing had changed. In the ten years 1870 to 1880 almost two million "buck skins" were exported from Natal alone; 485,786 of these in 1870–71. One firm alone exported 157,000 blesbuck and wildebeest skins in 1866. Perhaps it was this sort of scale prompted Parker Gillmore to write in 1876: "I did not wish to kill game, but to see it." Although even he did his share of wounding and slaughtering and of the giraffe he wrote: "Calf giraffe is excellent and nutritious food, so is fat cow, but save me from having to feed upon old bull".

Chapter 4
Requiem for the Rhino

Abstract Large numbers of black and white rhino recorded by early hunters throughput country. Several "species" described. Large numbers of both types killed. Extinction of white rhino predicted in 1855. Both almost extinct by end of nineteenth century and becoming so in the twentieth century. Black and white rhinos re-introduced in recent times.

Keywords Rhinoceros · Botswana · History · Quotations

4.1 Numerous from North to South

Both black and white rhinoceroses were once numerous in Botswana. From south to north along the eastern fringe of the Kalahari desert, westwards south of the Makgadikgadi Pans across to the Boteti, down to Ghanzi and north to the Chobe River. Somewhere well south of Makoba was a deep pit holding a little water called Bachukuru, or "place of the white rhinoceros". The former richness for rhinoceroses of the now waterless western extremity of Botswana is borne out by the name Tshukudu Pan, halfway between Ncojane and Mamuno; and the numbers which Galton and Andersson came across west of Ghanzi, as well as at Ghanzi itself; while McCabe found their tracks on the Okwa near Tswaane.

From the beginning of the nineteenth century travellers began making common acquaintance with rhinoceroses in the region of the Molopo to Marico rivers. Bain shot his first white rhinoceros near Lotlhakane in 1826 when the species was almost unknown to naturalists, later reporting four other rhinoceroses there. He also found tracks near Korwe, and in 1834 he hunted rhinoceroses along the Molopo near Phitshane and "bagged a good number", as well as catching six zebras. White rhinoceroses were found in plenty along the eastern border south of Gaborone, and westwards in the hilly country to Kanye and beyond. In May 1852 Chapman shot four rhinoceroses on the Notwane River, recording the sighting of at least a dozen on one day near Malotwana.

C. Spinage, *Big Game of Botswana*, https://doi.org/10.1007/978-3-031-60033-3_4

Cumming saw his first rhinoceros, "a huge white rhinoceros with a large calf", somewhere in the region of Mannyelanong. Half an hour later he fell in with a bull of the same species. Then, following along a stream, perhaps the Lotlakane, he came upon a bull black rhinoceros, or "borele" as he called it: "Both varieties of the black rhinoceros are extremely fierce and dangerous, and rush headlong and unprovoked at any object which attracts their attention." The white were "of a much milder and more inoffensive disposition… rarely charging their pursuer." Of their social behaviour: "they never associate in herds, but are met with singly or in pairs. In districts where they are abundant, from three to six may be found in company, and I once saw upwards of a dozen congregated together on some young grass, but such an occurrence is rare."

Black and white were found in the same areas, drinking at the same pools. They seem to have been particularly abundant to the north-west of Mochudi, as Cumming's experiences at a waterhole, probably on the Dikolakolana river, indicate: "At night I watched the water near my camp with Kleinboy [his attendant]. After a long time had elapsed an enormous old bull muchocho or white rhinoceros came slowly on, and commenced drinking within fifteen yards of us… Next came an old bull borele… Next came another borele… The fourth that came was another old bull muchocho… And fifth came a cow borele… Three other rhinoceroses came about me, but I was too drowsy to watch any longer, and fell asleep." But on the bank of the Limpopo River in the region of Martin's Drift in the Tuli Block, he counted "no less than twenty-two rhinoceroses, nine of which were in one herd, feeding on the open plain."

Cumming's most numerous encounters were in the hills to the north of Lobatse along the headwaters of the Notwane; to the north-west of Mochudi, and in the regions of the Shoshong and Tswapong Hills. They also occurred at a waterhole at Lephephe, today an arid area that one would not associate with the rhinoceros, and he also found an abundance of spoor on the Mahalatswe river near Mahalapye.

Methuen recorded seeing, and wounding, his first rhinoceros somewhere in the hills to the west or north-west of Lobtase. This was followed by two white rhinoceroses, a cow accompanied by a large calf; but "we have not yet encountered many rhinoceroses, but indications of their existence are numerous", he wrote. Still in the same area he came across two more, then found five sleeping together in the sandy bed of a dry stream.

All told he recorded seeing considerably more than 27, of which only one was identified as black and nine as white, one of the latter a calf. Most of the others were probably white also, and two were with calves. In this case the animals were seen in the space of three months, between August the 19th and October the 19th 1844. He hunted most of his rhinoceroses along the Marico River, south of Oliphant's Drift, but he recorded their tracks as numerous along the Notwane, probably in the region south of Gaborone (Fig. 4.1).

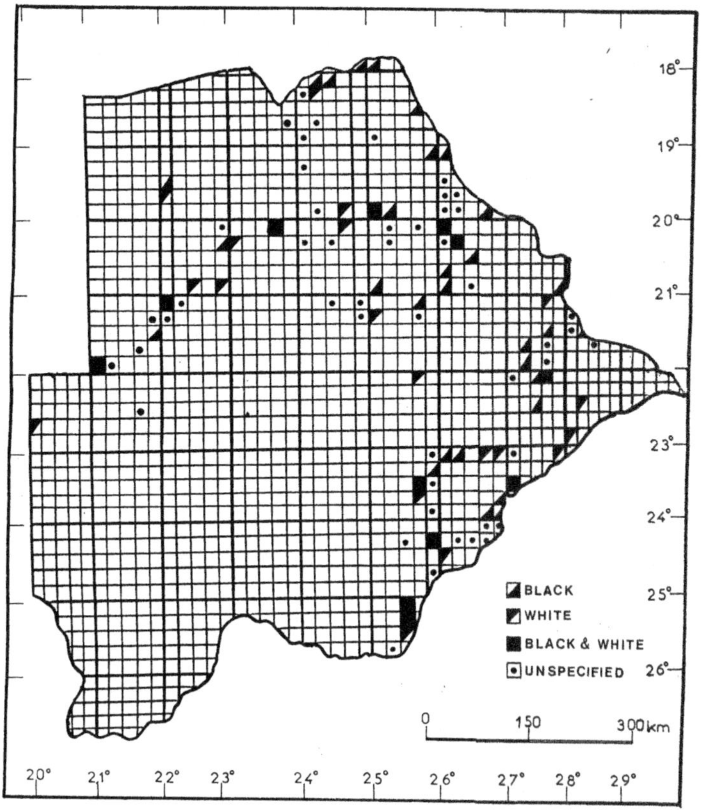

Fig. 4.1 Recorded presence of rhinoceroses 1824–1900. An unmarked square does not imply absence

4.2 How Many Species?

The white rhinoceros was first encountered at Chue Springs and recognized as a scientifically new form by William Burchell in October 1812. Chue Springs was in present-day Bophutatswana just south of the border at Werda, on a tributary of the Molopo. Rhinoceroses had been described from South Africa before this. Peter Kolb described the now extinct Cape black rhinoceros from personal experience in 1719: "If he makes at you, the way to avoid him is to suffer him to come within 8 or 10 Paces of you; and then to whip a few Paces aside; for then he loses Sight of you; and it costs him a great Deal of awkward Trouble to get you in his Eye again. This I have experienced my self; for he has more than once made with all his Fury towards me." His illustration however was simply a copy of Dürer's fanciful 1515 rendering of an Indian rhinoceros.

In 1743, Dr. Parsons, in a communication to the Royal Society, figured horns of 94 and 81 cm, which were probably those of the white rhinoceros, from Dr. Mead's and Sir Hans Sloane's private collections of curiosities. The species was described in more detail in 1775 by Dr. Andrew Sparrman, who visited the Cape in 1772–1776, and by Robert Jacob Gordon in 1778; but John Campbell seems to have started the controversy over the existence of different species in 1820. He recorded that two were shot which came into the 'town' of Mashow during the night, a kraal somewhere north of Vryburg, south of the Molopo river.

"They brought also the head of one of them, which was different from all the others that had been killed. The common African rhinoceros has a crooked horn resembling a cock's spur, which rises about nine or ten inches above the nose and inclines backward; immediately behind this is a short thick horn; but the head they brought had a straight horn projecting three feet from the forehead, about ten inches above the tip of the nose. The projection of this great horn very much resembles that of the fanciful unicorn in the British arms. It has a small thick horny substance, eight inches long, immediately behind it, which can hardly be observed on the animal at a distance of a hundred yards, and seems to be designed for keeping fast that which is penetrated by the long horn; so that this species of rhinoceros must appear really like a unicorn when running in the field. The head resembled in size a nine-gallon cask… it appears capable of overcoming any creature hitherto known". He added: "The animal is considered by naturalists, since the arrival of the skull in London, to be the unicorn of the ancients". The skull was deposited at the Missionary Museum, but its subsequent fate is unknown.

It was the missionary Robert Moffat, writing in 1842, who first referred to the supposed four different types. Mohuhu, the largest, he alleged had been known to kill an elephant by thrusting its horn into the elephant's ribs. This genus was called by the Tswanas chukuru. He claimed the four distinct species had been pointed out to him all within sight at once; the mohuhu, kheitlua, borila or kenengyane, and the name of the fourth he could not remember, but it was distinguished from the keitloa by the position of its ears and the formation of its head. The borele was the smallest and had the shortest horns but was the most fierce, the last to retire from inhabited regions. The other species, being more timid, sought the more remote places.

Cumming accepted there were four different forms, distinguished by the Bechuanas by the names of borele or black rhinoceros; the keitloa or two-horned black rhinoceros, first described by Doctor Andrew Smith based on a specimen obtained in the Transvaal; the muchocho or white rhinoceros, and the kobaoba or long-horned white rhinoceros.

Local names for the white rhinoceros were often used for the black after the extinction of the former. The modern Setswana name for the black rhinoceros is *tshukudu* or *bodile*; whereas the names for the white rhinoceros were *tshukudu*, *mogohu*, *kgetlwa*, and *kobaoba*. So different names may have been simply different Tswana dialects for the same thing; although the Kwena on the west bank of the Limpopo, told Dr. Smith there were two sorts of black rhinoceros; one with horns of equal length which never associated with the other, which had one horn shorter.

According to Cumming, the horn of the muchocho averaged two to three feet in length and pointed backwards, whereas that of the kobaoba often exceeded four feet in length and inclined forwards at an angle of 45 degrees. The posterior horn of both species seldom exceeded six or seven inches in length. Methuen stated: "We rarely found them with horns exceeding twenty inches in length, though specimens upwards of three feet long are occasionally obtained."

Livingstone reported the "kuabaoba" or "straight-horned rhinoceros… a variety of the white" along the Nhabe river. His theory was that as the horn projected downwards it did not obstruct the line of vision, so this species was able to be much more wary than the others. He noted the rhinoceros often had worms on the conjunctiva of the eyes (apparently a parasite which has never been identified), "but these are not the cause of the dimness of vision which will make him charge past a man who has wounded him, if he stands perfectly still… It probably arises from the horn being in the line of vision, for the variety known as kuabaoba, which has a straight horn directed downwards away from that line, possesses acute eyesight, and is much more wary."

Livingstone believed there to be only two varieties and the extra species were due to differences in size, age and the direction of the horns "as if we should reckon the short-horned cattle a different species from the Alderneys or the Highland breed."

"I find, however, that Dr. Smith, the best judge in these matters, is quite decided as to the propriety of the subdivision into three or four species. For common readers it is sufficient to remember that there are two well-defined species, that differ entirely in appearance and food."

In his private journal he wrote: "There seems after all to be only two kinds of Rhinoceros, the White and the black. Of the white there are the common Mohohu and Kuabaoba as varieties, and there is not a particle of difference in them except in the shape and direction of the horns. Of the black there is the Keitloa, or long horned variety, the two horns being of same length, and the borile, or little short horned variety which has not yet attained its full size. There is not a particle of difference between these varieties except in the matter of age and growth. The Keitloa is only the full grown black Rhinoceros. The colour of both white and black is the same."

Later he expounded upon this in greater detail, calling the white rhinoceros the "large long headed, grass feeding one." The variety kuabaoba was found in the company of the mohohoo and was in no respect different in feeding habits, etc. The chief food was grass, particularly *Cymbopogon plurinodis* and another species, but young shoots of some trees were also taken; such as acacias, Kalahari currant *Rhus tenuinervis*, raisin bush *Grewia bicolor*, karee *Rhus lancea*, and sodom apple *Solanum incanum*. The black rhinoceros was short-headed, feeding on trees and bushes, particularly acacias; *Catophractes alexandri* or trumpet thorn which still goes under the Tswana name of *mophuratshukudu*, or rhinoceros plant, and raisin bush *Grewia flava*; and it dug up a tuber. The variety boreelenyane, "Of the propriety of calling it such indeed I have some doubts", "or little black short-horned animal', fed on trees and bushes and its habitat was rocky hills. The keitloa was the full grown black rhinoceros.

The grain of the horn was finer in the black than the white. Livingstone noted that a white rhinoceros horn of 3 feet 10 inches in length "and another a foot longer", had been seen, but these were extreme cases. Three feet three inches and under was more usual. "The Kuaboaba is found in particular localities only and is scarce in all."

The white rhinoceros he noted was partially gregarious, occasionally being met with in groups of five to seven.

The white rhinoceros was the most inoffensive of all the large animals and never molested anyone, although the males sometimes broke in the frontal sinuses in their fights with one another. "As his flesh is usually loaded with fat, and a single bullet behind the shoulder slanting upwards a little in its direction is sufficient to kill it stone dead, it soon becomes extinct after the introduction of fire arms into a country."

Selous disputed the fact that it was entirely inoffensive and quoted several instances of the white rhinoceros attacking hunters.

Chapman did not believe there were two species of white, the kobaba and mohogu being only varieties, considering that in countries where guns had not yet been introduced the rhinoceroses were able to attain a greater age and hence had longer horns, which from growth and abrasion on the underside inclined downwards. In newly-opened countries long-horned rhinoceroses were always found at first which were selectively hunted for the length of their horns.

Elaborating upon this theme, Chapman believed there to be three varieties distinct from one another, besides one or two which were doubtful. First was the mohogu or white rhinoceros, also called the chukuru, mogoana and mogoshuana at different stages of maturity. Secondly there was the kobaoba (sometimes he calls it the kobaba) or long-horned white rhinoceros; and thirdly the khetloa or true and large borele or black rhinoceros. Then there was the borelengani or keningani, and the mogoshuana which he shot on the Botete river and which he believed to be another grass-eating rhinoceros and not a hybrid.

Chapman, while disbelieving in the existence of the kobaoba as a species himself, which had been given separate status by Dr. Andrew Smith, quotes experienced local hunters on the subject, who he points out are very observant of animals. They asserted it was more slender in body than the mohogu, with a decidedly hollow or saddle back and not quite so large, carrying its head and neck like a black rhinoceros although longer than the latter's. The horns were long and tapered forwards and the edges and insides of the ears were very hairy.

The borele, he noted, throws forward both its ears together and lifts them together. Feeding on bushes and roots it is nervous and fidgety when discovered, but confines its movements generally to the head, moving it from side to side undecidedly. The borelengani or keningani, sometimes apparently known as the chukuro bogali or fierce rhinoceros, had large ears, pointed and tufted which it cocked in a frisky way. A dumpy, plump-looking animal, always very nervous, always on the trot and often charging around even when it had not been discovered. A possible fifth variety, which he shot on the Boteti, which ate grass as well as browse, he does not describe; but his friend Frederick Green stated it to be a link between the black and the white. Although following the tracks of one on the Thaoge river he never came up with it. Green, writing in 1856, referred to local reports of a small black rhinoceros existing to the north of Lake Ngami which was not half the size of the other species.

The keitloa was generally described as having the two horns of equal length, an animal the Afrikaner hunters called the "blue" rhinoceros; whereas the common small black rhinoceros had a long anterior horn and a small posterior one. The long-horned white rhinoceros, or kobaoba, was so named as a separate species by Dr. Gray of the British Museum on the basis of a pair of horns obtained by Oswell, of which the anterior horn was straight and inclined forward at a right angle and the posterior horn was small. Just as in the "two-horned" black rhinoceros the size of the posterior horn is simply a reflection of the animal's age, probably the angle in the kobaoba developed with age and the weight of the horn, which often exceeded four feet in length; the backward-curving horn of the "common" white rhinoceros averaging only two or three feet in length.

Although Cumming alleged the kobaoba, "the rarer of the two", was found chiefly to the eastward of the Limpopo, long horns are characteristic of docile cows in saline areas, where they neither fight nor dig for salt. His one-horned rhinos were probably simply young ones in which the posterior horn had not yet grown. But horn growth can be very variable, unsupported as it is by a bony core; and there are both sexual and geographical differences in shape.

Selous wrote in 1881 he firmly believed there were only two rhinoceroses in all Africa and convincingly debunked the theory that more than two types existed.

Vague as they are, we can never now know whether any of the alleged differences were valid or not. But it is perfectly feasible for there to have once been distinct geographic varieties which had come together again, and some modern taxonomists incline to the view there was a certain amount of subspeciation. The now extinct Cape black rhinoceros *Diceros bicornis bicornis*, for example, differed from the race probably encountered in southern Botswana, *D.b. minor*; allegedly a smaller race than *D.b. chobiensis*, the race inhabiting the Chobe valley.

In spite of the thousands which were slaughtered, no skull of the white rhinoceros from Botswana is known to exist in any collection, so we cannot determine what races, if any, might have existed. Suffice to say the widely separated northern race of Zaïre, and formerly Uganda and adjacent countries, is almost indistinguishable from its southern counterpart.

4.3 The Massacre

Altogether Cumming records seeing, killing, or wounding, some 89 rhinoceroses, of which 30 were identified as black and 34 as white (probably the 22 were white as well making 56 in total). Twelve of the black are recorded as bulls and five as cows. Of the white rhinoceroses, ten are recorded as bulls and five as cows, with four of the cows being with calves. But when the time he spent hunting is taken into consideration, these occurrences comprising four trips in a period of about four years, the number of encounters is not as great as it at first appears. However, Livingstone recorded two other gentlemen hunting in the same region killing 78 in one season. Oswell and Vardon, perhaps the gentlemen referred to, killed 89 in one year. Oswell noted: "Vardon was the most enthusiastic rhinoceros hunter; he filled his waggons with horns as I did mine with ivory; he used to shoot four or five every day'. But apart from the sportsmen, the Afrikaner hunters often shot 50 or 60 rhinoceroses per man during a season. By the time Livingstone was writing in 1856 he noted sportsmen would not now find an equal number as in the 1840s "for as guns are introduced among the tribes all these fine animals melt away like snow in spring. In the more remote districts where fire-arms have not yet been introduced, with the single exception of the rhinoceros, the game is to be found in numbers much greater than Mr. Cumming ever saw."

In 1855, hunting along the Thaoge river, Chapman reported wounding several rhinoceroses, "most of which, as we afterwards learnt, had died." Despite this he wrote: "I believe that wherever guns are to be found at present the white rhinoceros is not allowed to reach its prime, and will soon be extinct." This was after he recounted how at Kanyu he shot a large bull white rhinoceros at a waterhole. After it fell 11 more appeared, two of them young ones. Having shot a cow, "I crept out and followed them, wounding successively three others". Two more were shot at before he went to sleep. The night before eight had come to drink but not within reach of him, but his companion shot at one.

"The rhinoceroses, all of which were white, occupied each twelve minutes to drink their fill, after which they wallow in the mud, or else go to their regular sleeping-places. At these their dung is found accumulated, sometimes to the amount of a ton or more. They like the warmth of the manure to lay in... The white rhinoceros likes the open plains, where there is just enough bush to shelter him from sun and wind. The borele likes the jungle of thorns (haakdorn), and the most secluded and retired spots. The khoetla, or large black rhinoceros, is more an inhabitant of rocky hills."

Kanyu seemed particularly rich in rhinoceroses, black as well as white were recorded visiting the waterhole. However, writing of the Makalamabedi area of the Boteti, he noted that rhinoceroses were "fast diminishing", and blamed the San.

In his trips, spanning a period of 11 years, Chapman lists at least 36 rhinoceroses as shot by him or his party, of which six were identified as black and eleven white; but the actual total was probably much higher. He also saw at least 28 others, on three occasions recording seeing "several". The Sowa river, Chapman noted, "hitherto unknown to white men", was famous for long-tailed rhinoceroses; perhaps he meant long-horned. When he got there, however, although rhinoceroses were numerous in the area east of Sowa pan, the river itself was a disappointment.

Just before Rietfontein, over the border west of Ghanzi in Namibia, in 1851 Galton had found rhinoceros skulls "lying in every direction, but strangely enough only one spoor could be seen". Forty had been killed there about a month before. He was told that the rhinoceroses would begin trooping in to the water at nightfall. Hardly surprisingly, they didn't. He next records spoor at the following camp and shot one, before meeting two white rhinoceroses near Rietfontein. At the Rietfontein waterhole itself the San reported rhinoceroses to be very numerous. As the only water within 16–24 km the number of animals that nightly congregated there was "truly astonishing".

Galton described the Hottentots as doing a great deal of mischief with rhinoceroses, for they lay six or seven together in a hide and blazed away volleys at distances of often 30–40 m, consequently killing very few, but many probably departed wounded and died elsewhere. Apart from shooting white rhinoceroses here, Galton also shot a black. Andersson, Galton's companion, records they shot upwards of 30 rhinoceroses, he alone killing eight in five hours which he claimed comprised three different species: "And it is my belief that if I had persevered I might have destroyed double the number. But I never took delight in useless slaughter."

Ghanzi was seemingly no different. In 1853 Andersson shot several there in a few days, on one occasion counting 20 filing to water. On his journey to Lake Ngami he shot almost 60. In that same year Livingstone recorded the black rhinoceros as remarkably scarce in the country north of the Zambezi and the white "quite extinct". "and will soon become unknown in the country to the south." This was due to its timid unsuspecting disposition rendering it an easy prey, "they are slaughtered without mercy on the introduction of firearms." The black rhinoceros he noted was possessed of a more savage nature, and, like the ill-natured in general, was never found with an ounce of fat in its body. "From its fierceness and wariness, it holds its place in a district much longer than its more timid and better conditioned neighbour."

On his first trip north in 1857, probably as far as Francistown where he then went east into Zimbabwe, recording his first at Lephephe, Baldwin saw at least 23 rhinoceroses in all, of which 19 were killed. Of this total at least four were black, including two cows and one large calf; and six were white. On his second trip he mentions only seeing more than eight, and killing one white. But the total was undoubtedly greater than this, as on his third trip he also only mentions seeing more than eight, but on his final trip in 1860 he records killing eleven white and twelve black.

On the Semowane river which flows into the east side of Sowa Pan, a drop spear apparently set for rhinoceroses, killed one of his oxen. The trap consisted of a poisoned barbed spearhead loosely inserted into a heavy 1.2 m length log of wood suspended in a tree and attached to a trip. Rhinoceros spoor was numerous in this area.

In the late 1860s William Finaughty shot 13 near to Plumtree on the border with Zimbabwe, and in 1868 the trader Thomas Leask brought to Shoshong almost a waggon load of horns which he had collected in Zimbabwe, an amount which would have weighed well over a ton.

Selous came across his first rhinoceros spoor in 1872 at the Motloutse river. Two years later he saw a black rhinoceros cow with calf on Puku Flats and came across another on the Chobe river, as well as hearing one drinking at night probably near Mabele. In August near Kataba he shot his first, a bull with an anterior horn measuring 2 feet 1 inch and posterior 1 foot 4 inch; and near Parakarungu he shot a cow with calf. Following along the edge of the swamp in the Kataba area he noted that a few rhinoceros spoors were crossed, not very old but "these latter animals are rather scarce all along the Chobe." Later in November he shot another bull at Tamasetsi Pan.

Hunting on the Nata River about 30 km from Sowa Pan in 1875, Frederick Barber wounded a black rhinoceros, "Had I known at the time how scarce rhino were, and how seldom found, should have let them go [giraffe] and stuck to the rhino" he commented.

The Portuguese Major Serpa Pinto, who crossed Africa from Benguella to Durban from 1877 to 1879, recorded that he saw his first rhinoceros at the Limpopo "where one would expect to find such animals scarce, on account of the war made upon them by the Afrikaners."

4.4 Some Freak of Fashion

It had been the custom for all male Tswana south of the Zambezi to undergo a circumcision ceremony or *boguera*, after which they could only marry when they had killed either a rhinoceros, an elephant or a buffalo. Nevertheless the animal was able to withstand this sustained offtake, only when a value was put upon its horn were its days numbered. Selous observed rhinoceroses were still very plentiful in southern Africa in 1872 to 1874, but considered between that period and 1890 both black and white were practically exterminated between the Limpopo and Zambezi rivers. In various years between 1874 and 1887 he came across black rhinoceroses (mostly in Zimbabwe) but never anything like the numbers he had met with in 1872 and '73. About 1880 rhinoceros horns attained a considerable commercial value, Selous thought "through some freak of fashion in knife-handles or combs or what not in Europe". By 1880 ivory had become scarce in southern Africa and traders in Zimbabwe for the first time began to employ local hunters to shoot rhinoceroses for the sake of their horns, no matter what length.

In 1881 he wrote that 20 years ago the white rhinoceros seemed to have been fairly plentiful in the western half of southern Africa, but was now (1881) extinct there unless it existed in the Caprivi Strip: "And this is not to be wondered at, when one reads the accounts in Andersson's and Chapman's books of their shooting as many as eight of these animals in one night, as they were drinking at a small water hole; for it must be remembered that these isolated water holes, at the end of the dry season, represented all the water to be found over an enormous extent of country, and that therefore all the rhinoceroses that in happier times were distributed over many hundreds of square miles were in times of drought dependent upon a single pool for their supply of water."

In 1877, during several months' hunting along the Chobe and along the Kataba swamps as far south as the Goha Hills, he only saw the spoor of two white rhinoceroses, although three years previously he had found them fairly plentiful in the same district. By 1879, in eight months spent covering the Boteti, Mababe, Mochaba, Savuti and upper Chobe rivers, he never saw a single spoor and all the San he met said that rhinoceroses were finished.

The black, like the white, he recorded had been almost exterminated in the more westerly parts of southern Africa. In 1879 there were still two or three drinking in the upper Chobe, to the north-west of the Savuti outlet, roughly the area in which they were to persist for the next 100 years. In the Caprivi Strip there were none, and according to local report never were any there, but H. Neale reported them to occur there in 1930. In 1890 Nicolls wrote that rhinoceroses were still 'numerous enough' in the Okavango swamps, but Bryden in the same year referred only to a few still lingering "in the swamps of the Chobe river… but they are very scarce."

The same author wrote in 1893: "here the hunters of a generation since… shot at them at night by the half dozen together. Oswell and Vardon in Livingstone's time slew eighty-nine rhinoceroses in one season; C. J. Andersson nearly sixty in a few months; and other hunters were constantly emulating these examples. As this sort of slaughter was practised by Dutch, Griqua, and native hunters also, it is not surprising that the rhinoceros, which even in daylight is not a difficult beast to bag, has all but vanished from the scene."

In 1899, seven white rhinoceroses were reported near Lake Ngami, and by 1903 these were down to four, according to Dr. Gunning of the Pretoria Zoo. Sometime after this they became extinct in Botswana but the black maintained a tenuous hold until 1992. A few black were reported somewhere west of the Okavango in 1913, and in 1925 Schwarz reported tracks on the road near Gcoverega which were presumably those of the black. A 1928 expedition along the Chobe saw rhinoceros tracks probably in the Kataba region, and in 1936 one was shot by a poacher in the north-west. Neale reported them as occurring on islands in the Okavango Delta in 1930.

Three were sighted on Chief's Island in the Moremi Game Reserve in 1962. Other sightings were made in the Mababe area in 1963–4, and near Gatgarra in 1970. In 1971 it was reported that a few still existed in the country, 20 being considered an over-estimation. One was seen in 1983 near the Savuti marsh and probably the same animal was photographed on the Savuti channel in 1985. Two sightings were made in 1989, but in 1991 probably again the same animal was the one which was killed on the Savuti channel. None were seen in an air search in 1992. Earlier sightings near Nunga and Nata may have been of animals which had wandered across the border from Zimbabwe, the species having been re-introduced into the Hwange National Park in 1961.

White rhinoceroses were also introduced into Hwange from South Africa, and a number of these wandered into Botswana, two dying near Odiakwe in 1968 when a pan dried up. Others were seen in 1968 on the Nata River. The previous year four were introduced, two bulls and two cows, into the Chobe National Park, which later produced a calf. In 1976 a further 52 were introduced to Chobe and the Moremi Game Reserve. In 1992 their numbers were estimated to be of the order of 27, extrapolating from a total of seven actually seen; having dropped from an estimated 100–150 in 1987. Four were captured in 1992 and moved to a sanctuary at Serowe but one later died of previous bullet wounds.

Two white rhinoceroses from Natal were introduced into an enclosure in the Gaborone Game Reserve in 1985, but one of these died.

A few might have lingered on in the north, but most believed the white rhinoceros had been eliminated in the wild. And after a tenuous existence for over 100 years in the face of the hunter's rapacity, it seems the black rhinoceros finally met the same fate that its white congener did in the century before.

In the year 2000 populations of both black and white were re-introduced into the Okavango from South Africa and most released in the Okavango Delta. These seem to have flourished and there was an estimated 400 in 2019, but in early 2023 it was reported that half of these had been killed by poaching or died of natural causes. The remainder were then captured, dehorned, and relocated to sanctuaries at secret locations.

Chapter 5
The Elephant in Botswana

Abstract Former wide distribution, scarce in some areas by beginning of nineteenth century. Large numbers shot. Destroyed by local people also. Greed for ivory. People in north learn value of ivory in mid nineteenth century. Livingstone viewed trade with encouragement. Numbers scarce before end of nineteenth century. Licences for non-native hunters introduced in 1893. Tusk size. Elephant races. The elephants of Chobe. Recovery of numbers. Problem elephants.

Keywords Elephant · Botswana · History · Quotations

5.1 Former Distribution

Like the rhinoceros, the elephant *Loxodonta africana* was once widely distributed in Botswana; along the Molopo and up the eastern side, reaching westwards right into the central Kalahari, covering the whole northern area and westwards to Ghanzi. In 1826 Bain expected to find elephants at Korwe to the west roughly halfway between Ramatlabana and Lobatse, although none appeared at the pan; and he saw many tracks in the region of Mosupa near Molepolole. In the extreme west in 1851 Galton reported elephants as common just over the border in the area of Rietfontein on the Buitsivango or upper Hanahai, now a dry watercourse. McCabe found recent tracks on the Okwa near Tswaane in July 1852, and Shelley's expedition found recent spoor at Kang and near Lehututu in September of that year. Chapman found fresh tracks in 1861 at Ghanzi, as well as solidified tracks in the limestone there. Elephant bones have been found at Hukuntsi Pan near Tshane in the south-west Kalahari, and at Kalkfontein pan north of the Okwa in the western Kalahari.

But by the beginning of the nineteenth century the elephant was already scarce on the Molopo. Cumming came across his first sign some miles to the north of Lephephe, but Methuen found the track of an elephant on the Notwane River probably somewhere south of Gaborone. Chapman found previous season's tracks in 1852 on the Notwane about 40 km north of Gaborone. One did not have to venture far north along the eastern border to still encounter them in some number, "Two marches" from

C. Spinage, *Big Game of Botswana*, https://doi.org/10.1007/978-3-031-60033-3_5

the junction of the Marico River, presumably with the Notwane where it becomes the Limpopo, Oswell found elephants in 1846 "in such large herds that we halted a week or ten days, and the ivory as it was brought in was piled up under my waggon."

Travelling up the Marico River the year before, Methuen encountered on the far bank a group of Afrikaners who had shot 71 elephants, 19 shortly beforehand. Alluding to the improvidence of shooting elephants with small tusks, the Afrikaners replied as usual that the supply was inexhaustible. In 1876 Gillmore came upon an Afrikaner encampment on the Limpopo at Zanzibar whose inmates had killed 11 elephants, 14 lions; giraffes and "innumerable" antelopes.

Somewhere in the region well west of Buffel's Drift, in May 1852 Chapman met with some 200 hundred elephants, mostly cows. By 1870, of the area where the Notwane joins the Limpopo, the German traveller Eduard Mohr wrote: "Elephants have now entirely disappeared from this neighbourhood, driven way by ceaseless persecution. It is quite a mistake to imagine them to have been all killed; indeed, I was assured by the well-known hunter Ziesmann that they emigrated in small troops, most of them following a north-easterly direction". In 1879 Gillmore came across over a dozen bulls south of this junction on the Marico river.

Cumming shot all of his elephants in the Kwena and Ngwato districts, never needing to go further north than Lose, less than 50 km north of Shoshong. But it was not until he reached Masowe Pan to the west of the Shoshong Hills that he met them: "I at once christened it 'the Elephant's own Fountain'. This was a very remarkable spot on the southern borders of endless elephant forests… Here and there lay a thick layer of soil upon the rock, and this was packed flat with the fresh spoor of elephants. Around the water's edge the very rock was worn down by the gigantic feet which for ages had trodden there… From the borders of the fountain a hundred well-trodden elephant footpaths led in every direction, like the radii of a circle… those leading to the northward and east [Shoshong Hills] were the most frequented, the country in those directions being well wooded."

By 1853 elephants had gone from the area north of the Shoshong Hills, "There were not, of course, the elephants now that were to be found during the time that Gordon Cumming rambled between these hills, but every other sport could still be found to perfection; occasionally, only, a troop of stray elephants visit these parts, and they are either killed or driven out of the country immediately." wrote Chapman. In 1890 Bryden recorded how "a few years since" a troop of 9 or 10, mostly cows and calves, turned up in the hills near Molepolole. Everyone turned out and they were quickly destroyed.

Mackenzie wrote in 1860: "Although a few come from the Kalahari annually in the dry season to drink at one or other of the waters between Sechele's and Sekome's, and although in the habitat of the tsetse east of Shoshong, they are still to be found, yet hunters who hope to fill their waggon with ivory during the hunting season are now found every year in the district of the Zambese and in the country of the Mashona, to the north-east of Moselekatse." In 1865 Thomas Leask was to write: "Ivory is very scarce, there being few or no elephants left in the vicinity [Shoshong]; they have all retreated further inland."

"Eight or ten days' from Lake Xau, presumably along the Boteti river, in 1846 Oswell came across a herd of at least 400 elephants "standing drowsily in the shade of the detached clumps of mimosa trees. As far as the eye could reach in a fairly open country there was nothing but elephants. I do not mean in serried masses, but in small groups." Along the Nhabe River Livingstone recorded elephants "in prodigious numbers" on the southern bank. But even these, so far inland, were to almost disappear within the space of little more than ten years.

5.2 The Greed for Ivory

The first missionary to reach the Ngwaketse in southern Tswana, the Reverend W. Edwards in 1807–8, immediately began by trading for ivory, eventually quitting the ministry and buying a farm in the Cape with his profits. On his visit in 1820 to Kaditshwene about 40 km east of Lobatse, the missionary Campbell found trade in ivory already established there. In exchange for a tusk the chief had obtained some beads from the Bakwena. The Bakwena obtained the beads from Portuguese sources on the south-east coast also in exchange for ivory. The people were greatly disappointed that Campbell had not brought beads to exchange for ivory.

"Sicomy [Sekgoma I] was in those days [1845] in the possession of very large quantities of splendid ivory, and still [1849] considerable quantities pass annually through his hands." wrote Cumming. "Since I first visited Bamangwato, and taught the natives the use of fire-arms, they have learnt to kill the elephant themselves; but previous to my arrival they were utterly incapable of subduing a full-grown elephant… All the ivory which Sicomy then possessed, and the majority of that which still passes through his hands, is obtained from elephants slain with assagais by an active and daring race of Bushmen inhabiting very remote regions to the northward and north-west of Bamangwato."

Altogether Cumming shot 107 elephants in four years, and by all accounts wounded a great many more.

It was not until after Livingstone reached the Linyanti-Chobe rivers that the people there learned of the value of ivory. In May 1849 he received information from the Tawana country of Lake Ngami of "'flaming accounts of the quantity of ivory to be found there" and that cattle pens were made of tusks of enormous size. When he reached there in July of that year he found the accounts largely true, and saw for himself 13 instances in which the tusks had been left to rot where an elephant had fallen. The Tawana called them bones, "but in less than two years after our discovery, not a man of them could be found who was not keenly alive to the great value of the article."

It was apparently not white traders who were the first to follow in Livingstone and Oswell's footsteps, but Ngwato. The trader Wilson accompanying Livingstone and Oswell exchanged the ivory at one musket worth 13 shillings for ten "good large tusks". These Ngwato traders exchanged something like 20 lead bullets for 36 kg of ivory, others 40 kg for a musket, but the Kololo would only exchange one bull and one cow tusk for each gun, an average of about 32 kg. Ivory then sold at the Cape at five shillings a half kg, and a secondhand musket ten shillings. But having discovered that they had been duped the Kololo went to the opposite extreme, demanding exorbitant rates of exchange. When Chapman got there in 1853 they had an abundance of ivory but he could not reach a sensible agreement on price and did not trade.

In 1851 Livingstone viewed with encouragement the development of the ivory trade on the Boteti since the period two years before when the people had regarded tusks as worthless, because he thought that a trade in such merchandise would eclipse the trade in slaves. Chiefs and others could get what they wanted by bartering ivory instead of people. Livingstone wrote in October 1851 that he knew of 900 elephants having been killed on the banks of the Boteti since his first visit in July 1849, and a trader in 1850 took 10,350 kg of ivory to Kuruman, "the greater portion of which came from this river. The supply of this one article cannot be expected to continue always so large, yet when one river alone is known to swell the commerce of the Colony, what may not be expected from the numerous rivers indicated in the sketch?" [a reference to a map].

In 1856 Chapman estimated that 2000–3000 elephants were being killed annually south of the latitude of Lake Ngami, a figure which may have included the Zoutpansberg region of the Transvaal. The largest portion of the ivory was exported via Natal and Port Elizabeth, and a small quantity occasionally from Cape Town.

At Molepolole in 1857, Sechele I traded 360 kg of ivory with Baldwin for a waggon, of which two of the tusks weighed together 69.3 kg. That was all Baldwin seemed to have got on that trip and the following year he planned to hunt for ivory at Lake Ngami. But in April, when he had reached the region of Letlhakane, south of Makgadikgadi, he remarked: "Elephants are, indeed, hard to come at now." Reaching the region of Lake Xau in May, he records the elephants had all trekked away from the river (Boteti) since the last rains, but eventually he found fresh spoor: "The elephants are wary, and very hard indeed to come at, as they are now so much sought for, and every savage knows the value of the ivory." In all he obtained about 315 kg on the trip.

But Chapman had reported in 1852 that the Galikwe or Dennasena San had "swept the country of all the elephants with their spears alone", referring to the area north of the Makgadikgadi Pans. Young men went in and hamstrung the elephants while others dodged in and out spearing them. This was done on foot, and many men were reportedly killed. Once having killed the elephants, everything except the entrails had to be forfeited to their Ngwato "masters". Despite their ravages, in the region north of the present Nata Ranches, in July Chapman came across a herd he estimated at 500 strong, chiefly cows and calves: "The trail extended a quarter of a mile in breadth, and hundreds of fine trees, recently broken down by the herd, lay across our path… we returned through dense forests, recently trodden by hundreds of elephants".

With Cumming, Oswell thought the Tswana were not very good elephant hunters, unlike the San in elephant country. The San would ambush an elephant, one throwing a spear into it. The elephant then charged the man whereupon another ran up behind as close as he could and blew a shrill whistle, generally made from the leg-bone of a crane, which he carried round his neck on a leather thong. At the sound the elephant stopped and turned, whereupon another spear was thrown and the thrower again charged. This was repeated until the elephant was exhausted, final blows being dealt by men running in and stabbing it behind the ribs while it was distracted from the front. In this manner a dozen San could kill two or three elephants from a herd, whereas Oswell had seen an elephant with forty spears in it thrown by Tswana and they still could not bring it to bay.

Leyland, however, in 1851 considered most of the elephants obtained by the natives along the Boteti and in the Lake Ngami region, were by pitfalls. He noted elephants were very numerous on one part of the river, probably the Makalamabedi region: "They feed within a radius of ten or twelve miles from the river, coming down at midnight to drink. Large branches and small trees, that had been torn up by the elephants, were strewn about in the forest". While he was here some Afrikaner hunters killed a bull with tusks weighing upwards of 36 kg each. Yet a day's journey before the Thamalakane junction he came to a Bushman kraal on the Boteti river where they did not know the value of ivory. Having caught an elephant in a pitfall they had left the tusks in the skull. Nor had this group seen a white man before.

In 1859 when at Nkokwane on the southern edge of Ntwetwe salt pan, Baldwin met the well known Afrikaner hunters John Viljoen and Piet Jacobs who had killed 93 elephants in the region of Mababe or Shinamba Hills, having come across elephants in great numbers. The elephants there "have never been fired at, and are perfectly tame and in great numbers." But he admitted he had no cause to complain for he himself had almost 2,250 kg of ivory, although he had earlier remarked at the end of June "Elephants are very scarce this year, and bad to find", and "it is certain that there are very few elephants here [north of Ntwetwe pan] this year". He found them farther west, possibly in the region of Nxai Pan and then towards the Thamalakane. One from the latter region had tusks of 31.5 kg each; another two together had tusks weighing 113 kg in all.

The following year he spent five days in the area of the Victoria Falls without finding elephants. He had better luck at Jare Pan when hunting in the border area, noting "Tusks are gradually accumulating." But it was hard work because the elephants stood so far off from water: "They have been persecuted this year much more than usual." Altogether on that trip he and his assistants shot 61.

In May 1862 Chapman wrote elephants were "coming again to drink at the Botletlie, now that it is neglected by Englishmen and Afrikaners, who cause the greatest destruction to them". Of the country around Tamafupa Pan in July 1862 he was to write: "Now, before we can reap any of the advantages, every elephant has been driven out of the country, which a few years ago swarmed with hundreds. Since those few we killed in the Lake country [Lake Ngami], we have not seen another."

After crossing the Shashe and Tati rivers in 1863, Mackenzie wrote: "we were now in the heart of what was until recently one of the most famous resorts of the elephant. Immense troops still visit this district every year... Near to the confluence of the Impakwe and Ramokwebane, there are some acacia thickets into which no hunter can follow the elephant, and towards which the sagacious creatures, when alarmed, at once direct their course". In 1872 Selous remarked this region had very recently been one of the best hunting-grounds for elephants to be found in southern Africa.

A timely warning was provided by the grave of an Englishman named Firming, killed by the first elephant he had ever seen, "but these animals, when much persecuted, soon shift their quarters, and now both elephants and hunters have left these parts" wrote Selous.

By 1870 or 1871 most elephants had retreated into tsetse fly country where it was impossible to follow with horses and oxen, so mounted hunting gave way to hunting on foot, its relative difficulties and dangers compensated for by the increased sophistication of firearms.

At Tati, Oates reported in 1873 the ground was "quite ploughed up" with the spoor of elephants which used to come there five years before, and had been found near since. In February 1874 a herd of at least 200 came close to Tati.

Bryden noted in 1890, presumably referring to these elephants, a strong troop was to be found "in some wild and almost impenetrable jungle country between the Zambesi and Linyanti roads. This troop is partially preserved, and the country where they have found shelter is so dense and thorny that they are now practically unmolested."

Near the junction of the Simukwe (in Zimbabwe) with the Shashe, in 1868 William Finaughty, having just chased a herd of "about 200" on the Zimbabwe side, came back to the Shashe where: "For at least a mile and a half the south bank was literally black with elephants. It was impossible accurately to estimate their numbers, but there were literally thousands". He and his companion shot 45 between them. Later he was told by Khama this herd assembled at the end of summer somewhere low down on the Limpopo when the fruits of the marula *Sclerocarya caffra* were ripe, elephants congregating there from all parts every year. But they had been shot at by some Ngwato hunters which had caused them to move up the Shashe.

In 1868 Leask noted the area of Matlamanyane to the north-east of the Nata Ranches, was a part of the country much frequented by elephants which, although hunted for many years, were always to be found in it owing to the bush being very dense.

5.3 An Occasional Troop yet Wanders

In July 1886 Westbeech was to write he had always considered the north bank of the Linyanti the best of the hunting veldt for elephants, "but now it is not worth the trouble of hunting it". According to native talk, since the veldt had been hunted by San for two years, and a number of hunters to whom permission to hunt had been given by Lewanika, the elephants had been driven through the river, presumably meaning that they had gone to the Botswana side, coming this side but never staying. Lewanika had also given Moremi permission to hunt the Linyanti. He had a group of hunters searching all directions about Linyanti, and not one of them had seen an elephant. Nevertheless he obtained 22 tusks and some of his hunters reported plenty of spoor on the Kwando.

In 1884 elephants were extremely scarce in Ngamiland, and a band of Moremi's hunters, which had killed ten near Nokaneng, reported that when a troop wandered into this country the natives turned out in hundreds to hunt them on behalf of the chief, to whom all ivory belonged. Nicolls reported in 1890 they were "not now very numerous" as they had been lately so much worried by Moremi's men who were armed with Martini-Henry rifles. Except for a few instances they had forsaken the country south of the Lake but were still "numerous enough" in the Okavango swamps. A troop or two still drank on the south bank of the Boteti, ranging from there to the north side of the Khwebe Hills; but by the mid-1890s this troop had been wiped out by Tawana hunters.

"Along the Chobe River swamps, and occasionally in the Mababe veldt, an occasional troop yet wanders, as I hear." wrote one hunter.

In 1875 on the north bank of the Chobe near Kazungula, Holub came across the graves of Basubia chiefs with elephants' tusks inserted into them root downwards. Some were bleached and cracked by exposure, but the smaller ones near the centre of the graves, weighing about 9 kg, were generally in a better state of preservation. Those which had been deposited most recently he considered to be only "milk-teeth". These were probably small tusks with the enamel cap still on, but nevertheless permanent tusks; their small size indicating awareness of the value of ivory.

Livingstone had found the same thing 22 years before. On the north bank of the Zambezi between Kazungula and the Victoria Falls, the grave of a chief Sekote was garnished with 70 large tusks, and 30 more were planted on the graves of his relatives. All were decaying, but a few in the shade were in fair condition. "The ivory gravestones soon rot away. Those of Moyara's father, who must have died not more than a dozen years ago, were crumbling into powder; and we found this to be generally the case all over the Batoka country.'"

In 1893 licences for elephant hunting by foreigners were introduced in the then Bechuanaland Protectorate. There is little record of hunting at this time, for the great days of the ivory scramble were already over. In 1927/8 714 kg were reported exported to South Africa; 274 kg in 1929/9; 960 kg in 1929; 487 kg in 1930; 383 kg in 1931, and the next record is 204 kg for 1936, with none the previous year. In 1931 it was reported so little ivory was exported it was not worthwhile considering raising the tax on it. In 1937 only 19 tusks are known to have been exported from the Tawana Reserve, the main elephant hunting area. After 1936 ivory exports were lumped with "Miscellaneous animal products" and can no longer be traced.

5.4 Tusk Size

In 1826 Bain was presented near Kamye with a tusk "of at least" 41 kg by the Ngwaketse chief Sebego, but its origin is not recorded. A "good many" other tusks were also bartered from him. When Chapman met with the herd of 200 elephants near Buffel's Drift in 1852, one bull was killed with tusks weighing about 23 kg each. Six or seven cows were later met with and one was shot with tusks weighing together 16.7 kg. Others in the party shot a bull with 23 kg tusks, another with 8 kg tusks, the whole totalling 124 kg. Another section of the party shot 12 cows whose tusks weighed altogether less than 60 kg, or just over 5 kg each tusk.

Finaughty's 316 tusks taken in eastern Botswana and western Zimbabwe between 1864 and 1870 had a mean weight of 11.25 kg each. Hunting from the Shashe River in 1869, 53 elephants were taken with tusks averaging just over 12.6 kg each, but one bull had tusks totalling 112.5 kg. Barber was to write in 1878, hunting over the border in Zimbabwe north-east of Tati, that they collected about 765 kg of ivory, the largest tusks weighing 24.75 kg each, but the average was about 7.65 kg per tusk "not a very good average, but when one considers the number of elephants that are yearly killed, every hunter picking out the biggest if possible, it was as good as we could expect". One of Leask's hunting party in Zimbabwe in 1868 "shot eleven cows and calves: his largest tusk weighs only fourteen pounds [6.3 kg]." In 1871 one Tom McMaster hunted the Mababe flats obtaining 540 kg of ivory, probably representing 15–25 elephants. In 1887 Alfred Bethell, Adjutant the Bechuana Border Police, wrote that the average tusk weight was 14 pounds (6.3 kg).

At Kedia Hill, south of Lake Xau, where elephants abounded "in immense numbers", Oswell shot an elephant with tusks weighing 47.7 and 44.6 kg. The former weight Livingstone believed had only been exceeded in one instance of a 61.2 kg tusk, the locality of which is not recorded. He considered that size increased as one approached the equator and later heard of a tusk of 71 kg obtained in Angola; but as tusk weight reflects the age of the elephant this was simply a reflection of hunting pressure.

Coillard reported that in 1877 the traders at Shoshong estimated more than 67,500 kg of ivory passed through, which he calculated to represent more than 12,000 elephants, or an average tusk weight of 2.8 kg each side, and commented: "At this rate, one need not be a prophet to predict the early extinction of these animals."

The trading party of Gerald McKiernan collected about 3182 kg at Lake Ngami in 1877, and had more at Ghanzi, mainly trading up the Thamalakane River. McKiernan reported that about 13,600 kg were obtained at the Lake Ngami post last year and this year would be as much or nearly so. A great deal of this went westwards through Namibia. While at Lake Ngami McKiernan met Van Zyl's party from Ghanzi which had killed 400 elephants carrying a weight of 3,636 kg of ivory, noting "most of them with light tusks". In fact they averaged only 4.5 kg each side.

Van Zyl went down in South African folklore for this party having killed 103 of the elephants in one day. Criticized on account of the small size of the tusks, it was claimed the herd got into a panic in a waterhole and the animals were trampling one another, as a result of which a large number of young ones had to be destroyed.

But large tusks were still to be had, and the Pandamatenga trader George Westbeech about 1882 allegedly obtained a pair from the Barotse chief Sepopa weighing 92.72 and 90 kg, which he sold on to the Shoshong traders Francis and Clarke. But this was not Botswana ivory although the Ngwato chief Sekgoma I claimed he had seen larger tusks than these.

About 1878 the annual export of ivory of the trader Taylor at Shoshong averaged £30,000 (equal to 36,000 kg at 1885 prices).

5.5 Elephant Races?

As with rhinoceroses, so with elephants hunters were always trying to identify different races. Bryden considered the drier the country, the smaller the elephants. On the Limpopo river he claimed that the average height of a bull was 3.35 m, but on the Boteti River and "through the Kalahari", it was 3 m. But it was Livingstone who first expressed this opinion. He claimed the elephants which he met with on the Nhabe River were smaller than in the countries further south: "At the Limpopo, for instance, they are upwards of twelve feet high [3.7 m]; here, only eleven [3.35 m]".

Expanding upon this, he wrote those in districts north of 20° of latitude were smaller than those existing south of this latitude, despite a greater abundance of food: "The first time that Mr. Oswell and myself saw full-grown male elephants on the river Zouga [Boteti], they seemed no larger than the females, (which are always smaller than the males,) we had met on the Limpopo. There they attain a height of upwards of 12 feet. At the Zouga the height of one I measured was 11 feet 4 inches, and in this district [southern Zambia] 9 feet 10 inches. There is, however, an increase in the size of the tusks as we approach the equator".

Chapman recorded that west of Ncojane, perhaps over the border in Namibia, some "elephants of the desert", or makolonkouns, had been found and forty killed. He described the elephants as "diminutive", feeding on melons and able to live without water. However the desert elephants of western Namibia are much bigger than other elephants, although their tusks are small. The hunters apparently had to abandon the tusks because they had no water, and nothing more seems to have been heard of the makolonkoun except Livingstone's account in his private journal, where he refers to a small variety which does not rise above 2.4 m, called *makongkwane*, the plural of *lekolonkwane* (*makolonkwane*).

In 1874 Holub reported a herd of "small-tusked elephants" inhabiting the region of the Serorome, which then carried water. Allegedly this herd haunted the area for another two years before it was destroyed by Afrikaners trekking to Damara.

5.6 The Elephants of Chobe

In the triangle formed between the confluent of the Zambezi and Chobe rivers, in the 1870s at the end of October an annual elephant hunt was conducted by the Barotse from Zambia, just before the rains when the rivers were lowest and the grass at its driest. The grass along the base of the triangle was fired and the elephants driven into its apex where they were surrounded by over 200 canoes in the rivers and about 2000 men. The chief and some 200 men faced them and fired, which was the general signal for everyone to fire or hurl spears at them. In this way up to 300 could be killed, although when witnessed by Captain Norman Macleod, he saw only 14 bulls, of which nine were killed.

Firing was begun at 100 m which could not have been very effective and the mass of elephants was usually allowed to break back and escape. In another hunt recorded by Holub in 1876, the Barotse chief, Sepopa, opened fire at 50 m, which immediately put the elephants to flight when, as Holub describes, "a sort of panic seized everybody, guns were fired in every direction, often without aim at all, and in the general pell-mell it was no great wonder that only five elephants should be killed altogether." On one day the chief's hunting party sighted more than 100 elephants in the swamps and fired at least 10,000 shots, killing only eight. The largest pair of tusks weighed some 54 kg; six tusks weighed between 11 and 14 kg, four were small cow tusks, and four were very small tusks of no value. The large tusks had been "much injured" by bullets.

When Selous first reached the Chobe River in 1874 he found elephants, but not apparently in large numbers. They appeared to be wary for on a subsequent safari he noted: "A little after midday we crossed the spoor of a large herd of elephants that had come down to the river to drink during the preceding night. As I knew from former experience, that the elephants about here seldom stood except at long distances from the river". In 1874 he was informed by a coloured hunter that elephants had been plentiful in the dense jungle covering the sand ridges south of the river the year before.

Not for Selous the romanticism of Livingstone, the island of Xakumba was a "horrid-looking place" (because of mosquitoes and fever). Travelling west along the southern bank of the Chobe he found a dense continuous jungle, interspersed with large forest trees, which came down in parts almost to the river, remnants of which can still be seen today. "This jungle-covered land rises in some places abruptly, in others in a gentle slope, leaving along the shore a margin of open ground (from ten to a hundred yards broad), covered with short grass, and formed, no doubt, of alluvial deposit… as we proceeded, traces of the presence of elephants and buffaloes became more and more frequent…

"As we advanced into the bush, we found the tracks of elephants and buffaloes crossing and recrossing it in all directions… About an hour's walk from our camp I crossed an open valley, running down at right angles to the river, between two sand ridges covered with dense jungle, down which ran several large game paths, leading to the water, and well trampled by buffaloes and elephants. From the great quantity of spoor-of all dates up to the preceding night-I had no doubt that this valley formed a favourite route to and from the river".

Probably this was the Nyangwe, outside of the present Chobe National Park in the now settled area of Kasane township; but similar to the Sidudu valley within the Park. Following up this valley the next day he shot a bull with tusks weighing nearly 27 kg each side. Some days later he shot a cow near Puku Flats with tusks "only weighing 9 lbs. [4 kg] apiece." Altogether he collected over 180 kg of ivory on this safari.

When some local natives told him they had seen a troop of ten bull elephants come out of the forest on the other side of the river and drink and bathe in the afternoon, Selous remarked: "Fancy elephants in this age so little disturbed as to drink in the daytime, and what a glorious sight it must have been to see these gigantic animals walking in the open with their slow majestic step." According to the natives no gun had ever been fired on the islands around Parakarungu.

Later, probably somewhere south of Kasinka, Selous tracked a herd from the water into an opening in the forest: "When I caught sight of the herd they were at least a mile off, and advancing slowly in dense black masses... This was one of the largest herds of elephants I had ever seen; I am afraid to say how many of them there were, but I think there must have been from 100 to 200 at least. Excepting a few young bulls... they were all cows and young animals".

"It seems dreadful to slaughter so many of these huge creatures merely for their tusks; for, if there are no Bushmen or other natives about, the carcases are abandoned to the hyaenas and vultures. But *il faut vivre* [one must live]. Ivory is the only thing obtainable in this country to defray the heavy expenses of hunting... It is true that within a week I had killed ten elephants; but from this date (the 8th of September) until the 20th of November), though I was hunting continually the entire time, and tramped over an enormous extent of wild and utterly uninhabited country, I only saw one more of these animals (a young bull) in all that time." Bryden noted in 1890 that although Selous found them "plentiful" along the Chobe in the early'seventies, very few now remained.

When he hunted from the Shinamba Hills in the south-east of the present park in July 1853, Chapman found elephants to be very numerous, coming to drink at the pan. To the east of the hills he met one herd of five hundred cows and young: "I found myself in an extensive mogonono [*Terminalia sericea*] field, swarming with elephants in every direction, crushing down the bushes with a sound resembling the roaring of waves". After he had hunted the elephants in the area for three weeks, they apparently all moved eastwards. At this time the pans were fast drying-up, but when he moved out in mid-August, travelling north-west he found plenty of fresh sign north of the Ngwezumba river (Commane). Of the Kavimba region he wrote that elephants were so numerous and troublesome that in the grain season the natives had to "battle with them" to drive them from their plantations.

Returning to Shinamba in January 1855 he reported: "After travelling [eastwards] 30 miles over ground that two years ago was covered with spoor, I fell in with nothing... I continued... till March... Our search for elephants continued without success. They seemed to have all migrated into the tsetse country." He eventually shot a bull on one of the pans along the border with Zimbabwe.

Although he supposed they had migrated north-west, a major movement is from west to east into the Hwange area of Zimbabwe in the dry season. But one cow elephant, radio-collared in the Shinamba area in March 1987, did move directly north-west to the Linyanti, while a cow marked to the east of Shinamba and a bull marked to the south-east, also moved north-west. Clearly these movements from the Shinamba area are of very long standing.

The Shinamba Hills consist of two small conical hills, the larger one in the east formerly known as Man Shinamba, the smaller Woman Shinamba; the district formerly inhabited by a tribe of San called the Kgabo because their totem was *kgabo*, the vervet monkey. Near to the eastern hill is the pool by which Chapman camped. By 1909 this dried up in the winter.

Despite the onslaught, elephants still frequented the north-eastern border at the beginning of last century. Hodson reported plenty of spoor at Tamasanka Pan in 1907 and at Hendrick's Pan in 1909. In 1925, just north of Zweizwe, Schwarz found tracks of herds of elephants moving northwards. At Chosoroga he reported that "great droves" of elephants had crossed the track in the night on their way to the Mababe (Savuti Marsh). North of the Goha Hills in 1928 the tracks of "troops" of elephants were reported in the mopane forest, and later a troop was heard crashing away by a motorized expedition.

In 1932 all game was protected in the Chobe district, protection lasting until 1943. It appears to have been fairly effective, as contemporary references indicate. The Government Veterinary Officer reported in 1935: "The natives say elephants and buffalo were common in the thick bush about Kataba and Kasinka last year. This has not been the case for many years. There is no doubt that game has very largely increased in the last few years. Elephants were in large herds along the Chobe River all last winter as they used to be in Selous' time-the seventies."

The Kasane Resident Magistrate was driven to inquiring whether he could shoot an elephant in self defence if needs be. He wrote: "At present it is most difficult to be able to plan with certainty to go from Kasane to Kazangula (*sic*) in the afternoon. I have on three or four occasions had to turn back because of elephants on the road. They are quite peaceful… They are not a menace but merely a nuisance".

"I got into Kasane… from Kachikau. There were elephants from Kabulabula to Kasane. They have dug the road to pieces and I could smell and hear them all the way".

Komana's forest (probably the present Chobe Forest Reserve) had earlier been referred to as "the breeding place of the elephants' by another Resident Magistrate. However, one man living along the Chobe in the late 1960s after a long absence, recalled that in 1933 there was a well-known herd of 20–25 head between Kazungula and Kasane, but that otherwise elephants were scarce along the river. This does not fit with the Resident Magistrate's reports, or was he making a fuss about only 25 elephants?

San in the area in the 1960s reported elephant had been unknown to those living around the source of the Ngwezumba for several generations until about 1945. Then within a single year the whole area to the north, as far as the Chobe, "filled" with elephants which came from the direction of Masame, moved towards Leshomo and then across to the Chobe west of Kasane. Between 1940 and 1945 no elephants were seen in the Kasane area, but after 1945 small groups appeared and numbers increased yearly. A European who moved to Serondella in 1946 (where in the 'fifties there was a population of over 1120 people), alleged that it was 1949 before he saw an elephant in the region. He recalled that the movement was from west to east, starting west of Kachikau and Ihaha.

These movements may be explained by the fact that the 1940s experienced one of Botswana's worst recorded droughts. The summer rainfall area along the Chobe river, as measured at Kasane, was very dry in the 1930s, from 1930 to 1954 well below the long term mean (66 years) of 677.5 mm; the years 1955 to 1981 being above average. For 1945 rainfall was 14.5% below the long term average, while the following year it was 17.8% above average, so this may have accounted for the alleged influx of elephants.

Little hunting seems to have followed the cancelling of protection in 1943, which was due to concern at the eastward spread of tsetse. In 1953 the Officer in Charge of the Francistown police noted no ivory had been exported in that year on permit and he suspected it was being smuggled out, for in Ngamiland 44 licences were issued by the tribal regent in 1952.

Following complaints from the Chobe Concessions, who were exploiting timber in the Chobe area in 1952, the compound manager was authorized to shoot "the bull elephant in each herd that was causing trouble in the Concession". It was proposed to withdraw the permission the following year because the manager went shooting unaccompanied by a member of the police and because he appeared to be selecting the best bulls, as the tusks were up to over 27 kg "which is big for this part of the world". He was recorded as "recently exporting"' 137 kg of ivory, but this would amount to only three or four elephants. It was decided there "should be no shooting of animals which allegedly do damage miles from anywhere in the forest of the Concession". Thus it appears there was little, if any, elephant hunting in the area.

Between 1947 and 1952 the 22,000 cattle population in the Kachikau Enclave and along the Chobe river front collapsed to 2000, due to outbreaks of trypanosomiasis and streptothricosis. The latter is a disease causing hair loss but normally is not fatal to an animal in good health, and was probably a secondary manifestation of the poor condition of the over-stocked animals. This population collapse opened up the area at a time of exploratory behaviour in good rainfall to occupation by elephants. Extensive flooding then took place in 1957 destroying arable lands and causing the evacuation of flood-plain villages, aiding further the process of elephant colonization.

In 1960 the Chobe Game Reserve was declared, and in 1967 it became a national park, following which local people living as far away as Gweta; the fringe of the Okavango Delta; Kachikau, and in the eastern Caprivi Strip; alleged the elephant population increased very rapidly, especially during the following ten years. In 1966 elephants were reported for the first time in areas where they had not been known for many years, as in the eastern Makgadikgadi. Elephants also apparently increased across the river in the Caprivi Strip, signified by a rapid rise from the end of 1962 in the number of "garden raiders" shot. Counts along the waterfront showed a build-up in the five years to 1971.

Above average rainfall from 1955 to 1961 may have accounted for westward movements from Zimbabwe's Hwange Park in the east. Such conditions are times of exploratory behaviour and in the wet year of 1977 an elephant had the temerity to appear near Ghanzi but then disappeared. In a March 1984 wet spell one appeared near Gyom, south-west of Kedia Hill. Subsequently it re-appeared in the southern Kalahari where it was shot. Some young bulls moved into the northern Central Kgalagadi Game Reserve in the good rains of 1986.

The drought of the 1980s seems to have had no effect upon the population numbers, although the total amount of rain recorded falling at Kasane from 1978 to 1984 was 20% less than that recorded in the drought from 1929 to 1935.

Plate 5.1 The Chobe River is the elephants' lifeline

The elephant population in this area survived the massacres of the nineteenth century in relatively good numbers because it could not be hunted in the swamps and woodlands on horseback on account of the presence of the tsetse fly. Nor could waggons be brought into the areas, and all hunting and carrying out of the ivory had to be on foot. It has probably been increasing relatively undisturbed since about 1914. Hunting increased in the'sixties, and from 1979 to 1982 a total of 1,515 licences was issued. Probably over 500 licences a year had been issued prior to 1979.

Hunting, which accounted for about 1.5% of the population annually, was stopped in 1983 because of an alleged decline in tusk weight, but there was no statistical evidence for this. The average weight for 1979 to 1982 was 17 kg, the mean largest tusk weight being 33.8 kg and the absolute largest tusk weight 39.6 kg (1981). A "trade sample" of 2375 tusks imported into Hong Kong between 1974 and 1978, had a mean weight of 14.05 kg, a very high mean weight for a trade sample but probably reflecting the issue of citizen licences, the majority of elephant hunting being essentially "commercial". The mean weight of 65 tusks taken by American sport hunters between 1968 and 1978 was 22.42 kg, but this would be the result of very selective hunting (Plate 5.1).

5.7 Elephant "Damage"

By 1963 it was considered elephants in northern Botswana were exerting a destructive effect on woody vegetation along the Chobe River riparian strip. Cumming had first described elephant "damage" to trees in the 1840s, following the tracks of one bull into the hills: "Almost every tree had half its branches broken short by them, and at every hundred yards I came across entire trees, and these the largest in the forest, uprooted clean out of the ground, or broken short across their stems. I observed several large trees placed in an inverted position, having their roots uppermost in the air."

Methuen first saw signs of the effects of elephants on trees in the hills to the west of Mochudi: "acacias and other trees; their branches stripped of their bark by elephants to a height of fifteen feet. I have heard that these immense animals will, like goats, stand on their hind-legs with the fore ones resting on the trunk of a tree, and so reach the upper branches. They use their tusks as crow-bars in uprooting trees".

Particularly affected along the Chobe waterfront were mature *Acacia erioloba*, huge stately spreading camel thorn trees killed by ring-barking. From 1973 to 1975 it was estimated over 5700 elephants occupied the 11,000 km^2 Chobe National Park. In 1987 the number was estimated at 12,200. For an area of 22,500 km^2 in 1975 there were an estimated 12,000, where 17,800 were estimated in 1983. Beginning in 1980, aerial surveys suggested a population of 39,500 elephants, using a total area of 80,000 km^2 and depending on the dry season lifelines of the Kwando, Linyanti and Chobe rivers, which form a single tributary of the Zambezi. By 1989 the number had risen to almost 60,000, a mean estimated annual rate of increase of five percent, which is possible by reproduction alone without immigration.

West of Chobe along the Linyanti river, overall elephant density reached up to 4/km^2 in the dry season. Three times this density, comparable to that found along the River Nile in Uganda (12.5/km^2 in the 1960s), was recorded in 1000 km^2 in the Linyanti area in 1987. Much destruction of the riverine woodland was evident but these fine camel thorn woodlands may reflect the reduction of elephants in the nineteenth century, which allowed the trees to mature. By the 1980s they were ageing and more susceptible to over-use by elephants and to wind throw: numbers are blown down in wind storms or their branches broken off. Elephants may only contribute to the inevitable in aiding the disappearance of these mature woodlands, for large numbers of camel thorn trees were also dying where there were no elephants. Unfortunately early travellers tell us little about the riverine woodland, other than Selous's comments on the region of Kasane. Only Schulz and Hammar in 1885, having travelled along the Chobe and Linyanti south banks, across the Caprivi strip and up the Okavango and the Thaoge, reported at about the region of Gumare: "This day we passed a fine mimosa forest, composed of by far the largest trees we had ever seen; usually much smaller, many of them were much over thirty feet high". Which suggests such large trees had not been met with previously along the Chobe and Linyanti rivers.

The destruction of trees close to the water front near Chobe, first noted in 1963, may have been due to compression of elephants along this sector resulting from settlement to the east of it, military operations in the Caprivi Strip on the northern bank of the river forcing elephants across to the Botswana side in the 1970s, and immigration from war-torn Angola. Thus a complex of compression, protection, reproduction, and possible immigration, may all have contributed to an increase in numbers and consequent pressure on the riverine vegetation.

Total numbers fluctuate according to season, for an estimated 10,000 elephants move into Zimbabwe in the dry season. But a similar annual percentage increase has been experienced in the contiguous Zimbabwe population, in spite of a culling programme.

5.8 Theories of Increase

From a hypothetical model of the population increase, changes in vegetation with the destruction of mature trees may have started in the early 1960s, at a density approaching 0.2/km^2; which fits with the observation of tree 'damage' in 1963. It was postulated in Zimbabwe that specified woodland species would persist at an elephant density of 0.5/km^2, and closed canopy at 0.25/km^2. Tree regeneration and growth depends upon rainfall and, in its younger stages, fire; with the added complication of frost in some areas. Droughts will considerably retard tree growth and we may assume that periods of good growth have corresponded to peaks in rainfall in the area.

Destruction of mature trees and loss of canopy cover in the Chobe-Linyanti riparian strip was irreversible unless the entire elephant population was reduced to around 8000. Such reduction is unthinkable and it is arguable whether trees are more desirable than elephants, since there is extensive closed canopy woodland away from the waterfront.

Left to increase it is unlikely that a Tsavo-type die-off would occur, in which an estimated 6000 elephants died in one drought year in Kenya's Tsavo National Park. This is because the elephants at Chobe and Linyanti are probably supported in large measure during the dry season by the extensive flood plain grasses, which remain green throughout. Increasing competition for this food resource would more likely lead to a decline in reproductive success and a slowing of population growth. In Tsavo's semi-arid ecosystem there was no food resource close to the permanent water.

The elephants could disperse to the south and south-west of the Okavango Delta, where the density is presently very low. It is not known why the Delta is underutilized by elephants, although hunting was encouraged in the east, close to Maun, from 1939, followed by organized game extermination from 1942 to 1967 in the name of tsetse control. But elephants are not recorded in this area, nor is there any record of any being shot in the tsetse control operations, but the disturbance factor may have kept them away. Had they been present it would not have been possible to construct the tsetse control fences, the first of which was erected in 1944.

5.9 The Savuti Story

At Savuti, in the southwest of the park, a different situation pertains to that found along the Chobe-Linyanti rivers. In the 1980s a marsh only in name, fed by an intermittent channel, the Savuti Marsh is an area of about 100 km^2 of open grassland in the centre, bordered in the west by woodland of mature camel thorn and Kalahari Sand acacia *A. luederitzii*. The trees were characterized by being uniformly mature, evidently having grown up after a drying-out of the marsh in the late 1880s. Many were then drowned by flooding of the marsh in 1979, and in the 1980s the survivors were dying due to dryness. This resulted in the estimated 3000 elephants from the Linyanti region which used the area, together with about another surrounding 100 km^2 mostly in the dry season, exerting increased pressure on the remaining trees on the periphery of the former marsh area, hastening their destruction by ring-barking. The diminishing number of trees concentrates the elephants on the survivors, trees being attacked in proportion to their abundance.

Hence the forest of dead trees seen in the 1980s was only partially attributable to elephants. Regeneration, it seems, would take place only in a much wetter groundwater phase. A large area of candle-pod acacia *A. hebeclada* scrub in the south is not touched by elephants at all, possibly because the clay is riddled with treacherous sink holes. Selous described this same area in 1876: “The ground was one mass of holes and deep cracks, all overgrown with short grass”. Presumably due to a particular property of the clay.

The intermittent flow of the Savuti Channel is believed due to tectonic activity. It flowed again in 2010 flooding the marsh, and then ceased again in 2016.

5.10 Problem Elephants

In 1941 a remnant herd of elephants alleged to number 40–50 existed in the extreme east of the country, ranging between there and over the border into the Transvaal. General Smuts, when Prime Minister of South Africa, created a sanctuary for them on the South African side in that year, requesting the Bechuanaland Protectorate Government to do the same on its side. The Chief, Tshekedi Khama, on whose tribal land they occurred in Bechuanaland, refused to consider the idea of a sanctuary, replying that the elephants were already protected and only shot when they damaged plantations. General Smuts urged the Government to press the point fearing the land would be sold to farmers, but Tshekedi refused to reconsider his decision. The outcome was the sanctuary on the South African side was degazetted for farmland in 1949 and the farmers hunted the elephants, causing them to retreat to Zimbabwe and Botswana. Control operations in Zimbabwe forced the elephants back into the Transvaal where they were again hunted, concentrating them in the Botswana border regions which became elephant extermination areas, and a Game Control Unit was set up in 1956 to operate in them in the Tati and Tuli Block areas. This unit eventually became Botswana's Department of Wildlife and National Parks.

In 1959 it was estimated there were about 200 elephants in the Tati region. Five had been seen crossing the railway line near Serule to the south, and a herd of one hundred was said to occur in the Northern Tuli Block, which it was alleged moved seasonally to the north and back again. A remnant of these elephants still survives along the border with Zimbabwe, regularly breaking the border veterinary fence.

Years of harassment with no protected area to retire to made them furtive and aggressive, as they marauded through a relatively densely settled area, disappearing whenever attempts were made to track them down. The following extracts from a resident's 1989 letters speak eloquently for themselves: "I will deal with my home village, which is Tobane. At Tobane in the past elephants found a home belonging to a resident east of Tobane, these elephants brought her house down at her cattle-post and threw out her belongings.

"At our cattle-post at Tshutshumane, the same elephants found the herd-boy's bag of sour milk, brought the sack down and spread milk all over, unprovoked by anyone.

"After the disastrous drought, the elephants found a field with good harvest and devoured all the produce.

"At my late uncle's place they brought down the kraal and almost brought down the hut with some property.

"The latest incident, these elephants came to our cattle-post and brought down the kraal and [the] calves were panic stricken and fled into the bush.

"... the animals after doing all havoc and being satisfied, left for an unknown place.

"... everybody after the effects of drought did all to plough, only to feed the aggressive elephants which ate all [the] produce and went into residence[s] to devour what was kept there..."

5.11 The Tuli Block Elephants

Not far to the east of where this rogue herd operates, was an estimated 550–600 elephants in the Tuli Block, concentrated mainly in the private Mashatu Game Reserve at the junction of the Shashe and Limpopo rivers, but ranging over about 300 km^2. Although some assert this was all one population with the rogue elephants referred to above, the behaviour of the rogue elephants is quite different to those of Mashatu, and a connection therefore seems unlikely.

Today, protected in a private game reserve, this population exerted a considerable destructive effect on the riparian woodland, and many large mature trees were killed by ring-barking, no regeneration taking place. With an estimated total of about 590 in the 1980s the density was about 0.6/km^2. There was some suggestion the population has remained much the same size in numbers since 1976, but these elephants come into increasing conflict with surrounding farmland.

Although partially restrained by electric fencing in some areas, this can only be a temporary solution unless the population reacts rapidly with a decreased rate of recruitment due to shortage of food, as perhaps it may now be doing. However, to reach the total of 590 recorded in 1976, without immigration the population would have had to number around 250 head in 1940. Since elephant populations have often been underestimated by a factor of ten, it is quite possible there were five times as many elephants in 1940 as was then thought to exist.

The population appeared to be stabilizing at the relatively low density of 0.6/km^2, perhaps due to a scarcity of perennial grass in the area. Formerly heavily overgrazed by cattle, livestock rearing became uneconomical, and the area was turned over to hunting before becoming a private game reserve. But the entire populations of wildebeest and zebra died in the 1980s drought due to lack of grazing. The elephants therefore depended almost wholly upon browse. But good rains brought the grass back and the population rose to 900 by 2020.

5.12 A Capacity to Survive

Showing a remarkable capacity to survive in the face of man's merciless onslaught, Botswana's elephants have confounded the dire predictions of imminent extinction in the 1890s to go on from strength to strength. Up until 1991 the population in northern Botswana was one of the few remaining populations in Africa still increasing in the absence of poaching, and with more than six percent of calves in the population momentarily growing in numbers at its near-maximum rate. Limitations of habitat would eventually have come to bear on this rate of increase at density levels that we may suppose have been witnessed elsewhere in Africa, but as elsewhere the population would age and continue to exert its effects upon the habitat for many years to come. This appears to have been averted in Botswana, for with the cessation of hostilities in southern Angola and the lack of military presence in the Caprivi Strip, elephants dispersed into these areas and the indications are that the population level appeared to have stabilized, and in 2014 was estimated at about 130,000.

We can consider ourselves fortunate indeed that Bryden's pessimistic prediction of 1903 did not come true: "Such then, has been the rate of extermination in South Africa that the wild elephant has now practically ceased to exist south of the Cunene and Zambesi Rivers… In Khama's wide country of Bamangwato, North Bechuanaland, one troop yet remains, or remained a year or two since… If, fifty years ago, great Britain had suddenly extended her sway from the Cape to the Zambesi, the elephant might have been saved. Nothing else could have preserved it from extermination" (Fig. 5.1).

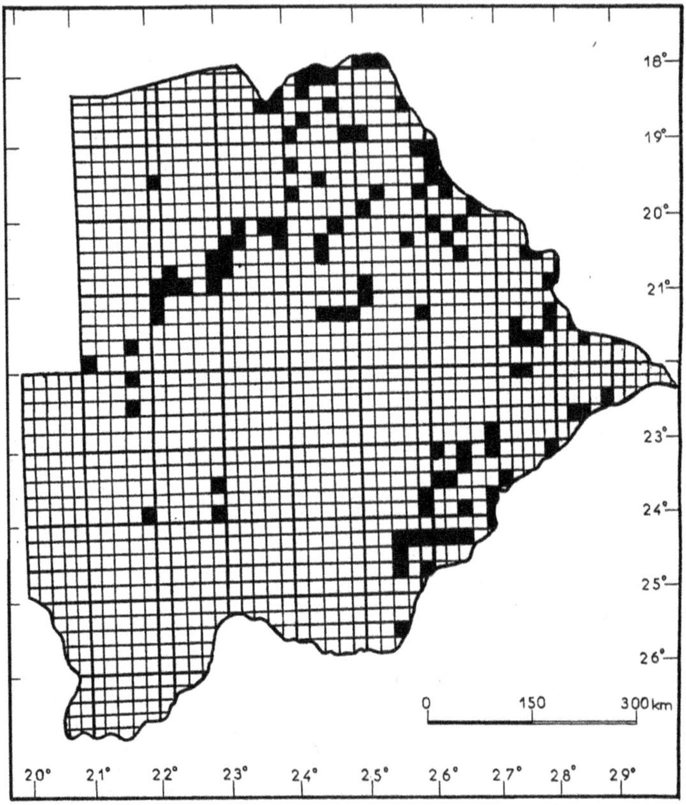

Fig. 5.1 Recorded presence of elephants 1845–1900. An unmarked square does not imply absence

Chapter 6
Demise of the Kalahari's Great Herds

Abstract Kalahari herds' movements, wildebeest population compared with Serengeti's. Decline of numbers 1960–1990. Deaths due to drought in 1980s. Movement patterns blocked by fences. Movements opportunistic and not set migrations. Comparison with Serengeti. 1930s mass die-off reported. Veterinary cordon fences exacerbate deaths after mid twentieth century. 1941 reports. Destruction of wildebeest and other species for meat and skins. Springbok and its irruptions. Other species declines. A fragile resource threatened by recurrent drought. The tragedy continues.

Keywords Big game · Botswana · History · Kalahari

6.1 Ordeal of the Wildebeest

"At last the lake was in sight, a miracle of blue waters, stippled pink and white by thousands of flamingos and pelicans. But even as the wildebeest surged forward, a new threat lay in wait. Out of the dust raced bands of poachers from the local kraals. Some had packs of dogs. Others used trucks as battering rams to run down their victims.

'The wildebeest were trapped… Soon the lakeshore and surrounding plains were littered with carcasses. When, at last, the rains broke and the herds departed, they left behind them at least 15,000 dead wildebeest.

"… in 1983, the tragedy was re-enacted on an even more horrendous scale. This time, when the wildebeest arrived, the drought had reduced Lake Xau to dust…

"By the end of September that year, Mopipi was a dustbowl landscape littered with thousands of carcasses. Animals which had withstood the drought were now too weak to escape the hunters, who pursued them with dogs or stoned them to death. A survey in December counted 50,000 dead wildebeest; and the stench was so great that at one point attempts were made to collect the bodies and burn them in a giant bonfire." (Jackman & Godwin, 1986).

The Kalahari desert is characterized by great herds of wildebeests, hartebeests, elands and springboks, which undertake nomadic movements in response to dryness

C. Spinage, *Big Game of Botswana*, https://doi.org/10.1007/978-3-031-60033-3_6

Plate 6.1 Wildebeest declined in numbers by 90% in the 1980s

rather than regular seasonal migrations. Most numerous probably since the turn of the century until the 1980s was the wildebeest, but even so its numbers were only a fraction of those occurring in Tanzania's Serengeti National Park. With a mean annual rainfall of some 780 mm, more than twice that of the central Kalahari, the Serengeti had a population in 1976 of 1.2 million and still increasing. In terms of supporting animal numbers the Serengeti is 13 times more productive than the near-arid Kalahari. Biomass or total live weight of game in the southern Kalahari peaked around 1979 after a run of good years at about 400 kg/km^2. This compares with an increasing biomass in the Serengeti in the 1970s of some 5200 kg/km^2. Put another way, the 51,800 km^2 Central Kgalagadi Game Reserve is roughly equivalent to a 4000 km^2 game reserve in Tanzania (Plate 6.1).

But we read little of the Kalahari's great herds in early accounts, probably because people travelled mostly in the wet season when game would be scattered, rather than in the dry when they might have encountered a migration. One of the few records is provided by Shelley's hunting expedition which in September 1852 encountered "vast quantities" of game between Dutlwe and Lehututu "which appeared to be generally trekking north-eastwards". In July of the same year Chapman reported "many thousands" of wildebeests and zebras in the region of Gutsha Pan, an area where annual concentrations still take place. And in the 1870s Selous observed the wildebeest congregated on the Mababe flats (Savuti Marsh) in immense herds during the dry season, as they do today. He only noted that wildebeests were found all along the eastern edge of the Kalahari desert because he never entered the central Kalahari himself.

At the beginning of the 1960s it was considered Botswana probably had the largest herds of wildebeests known in Africa, and estimates of 25,000 to 30,000 in a single herd were not uncommon. Previously the most numerous herbivore in the Kalahari, by 1990 its numbers had fallen from well over a third of a million in 1979 to little more than 32,000 for the whole country.

In 1980, at the beginning of the long drought period which ran from 1979 to 1987, in the first year an estimated 15,000 died at Lake Xau in the north-east of the Kalahari Desert. This was followed in 1983 by an estimated 50,000, and possibly as many as 80,000, more deaths. They died, not because there was no water in the lake, but because grass in the vicinity had been shorn bare by cattle. The wildebeest were faced with the choice of either dying of thirst where dry grazing was plentiful in the northern Central Kgalagadi Game Reserve, prevented from moving farther north by a veterinary cordon fence, or seeking water 40 km away and dying of starvation and dehydration-related imbalance. They chose the latter. In 1979 Lake Xau still had water with pelicans and flamingos sporting on its surface, and some wildebeest even died in the water indicating death was probably caused by "osmotic shock"; the disruption of the animal's physiology by a sudden intake of water when dehydrated, although it is alleged the nature of their stomachs enables ruminants to withstand this.

Four years later Lake Xau was dry, and the wildebeest trekked on a further twenty km to the Mopipi Dam where they could only drink at night, prevented during the day by people. Fifty thousand died around it of starvation, heat stress and probably osmotic shock.

But this was not the only place where wildebeest died. Some 250 km to the south, from Lephephe to Dibete, thousands of wildebeest as well as hartebeest died in 1980. Hundreds of carcases were reported as lying in Sojwe Pan. Along the Nosop river bed in the extreme south, although only a small part of the Kalahari population, relatively large numbers died, despite artificial water supplies which have been there since the 1930s. Here at least starvation must have been the cause of death.

Subsequent aerial survey estimates made in 1986 onwards to 1990 suggested more than 90% mortality may have taken place among the wildebeest in the Kalahari population, estimated to have numbered almost 330,000 in 1979. This estimate was made at the peak of a run of good years and just prior to the drought, although 1979 was already an exceptionally dry year and significant redistributions of species took place.

Such drastic population crashes have been a recurrent phenomenon in Botswana (and no doubt at one time in neighbouring South Africa), resulting from a 20-year drought cycle. But the former natural pattern of migration that was available to the populations in time of drought, with access to three river systems: the Orange, the Limpopo and the Okavango; all of which have now been cut off by settlement and veterinary cordon fences, does suggest that, under natural conditions, periodic drought mortality may not have been as drastic as is now experienced. It has also been suggested that much of the mortality may have resulted from habitat changes due to land mis-use. Perennial grasses have been suppressed by too-frequent fires and, around the central Kalahari, from overgrazing by cattle. In many places this has resulted in the invasion of former open grassland plains by dense thorny thickets, useful only to browsing species.

A major migration of the wildebeest has not been shown to be an annual occurrence, migration taking place only when the quality of the habitat falls below a certain minimum requirement, probably related to the amount of water in the food. When this happens the movement of the wildebeest is from a concentration area characterized by thousands of pans stretching in a broad band across the western middle of the southern Kalahari which break the monotony of this otherwise featureless sand plain. This is the "Bakalahari *schwelle*" or "rise"; the ancient watershed between the dead Okwa River drainage to the north and the Nosop and Molopo rivers to the south.

Concentrating together in dry years at the end of the normally wet season, from here the wildebeests move north to north-east covering as much as 50 km each night, a possible former northern migration to Lake Ngami or the Okavango Delta severed since 1958 by the east–west veterinary cordon fence which channels them eastwards to the region of Lake Xau. Here they arrive at the end of July and remain until October–November when the rains begin again.

Although after rain falls there is water in pans and good grazing in the northern Central Kgalagadi Game Reserve, the wildebeest still perversely migrate back to the south-west again as soon as conditions permit, but in scattered formation. They may do this at a time when calving is imminent to exploit the high levels of sodium and chloride present in some of the pans, many of these pans being richer in salts than are those in the north. Aggregation at the end of the rains is not related to breeding as concentrations in some animals are, for the wildebeests in the Kalahari mate in the rains about January when the herds are scattered, and calving takes place at the beginning of the rains about October. Not all of the wildebeests take part in the northwards migration, and it is thought pockets which remain may be relying on patches of tsama melons and desert cucumbers for their water requirements.

There are thought to be three major populations: one dependent upon the Linyanti River, moving between there and the Mababe area; another north of the Makgadikgadi Pans dependant upon the Boteti River; and the Kalahari population which occupies about half of the country and at one time extended into Namibia and the northern Orange Free State of South Africa. It is the latter population which has a northerly movement as the dry season advances. The extent of the movement and size of the concentrations which may develop appear related to the severity of a particular dry season and so are not repeated every year.

The mean annual rainfall gradient increases from south to north but more rapidly on the western side of the country, therefore we would expect the wildebeests to move north towards Ghanzi from the *schwelle*, which is exactly what they appear to do but they are channelled eastwards by fences. They then return in a south-west direction in the shortest possible line to the area west of Tshane because this is wetter than the area to the east. Farther south beyond the *schwelle* conditions are poor for plant growth. They are somewhat better on the *schwelle* itself but even better still farther north. The wildebeest thus move as far south as conditions permit.

Based upon four flights in a little over one year, it was shown in 1979 that under non-drought conditions the wildebeest occupied and vacated large tracts of the Kalahari seemingly at random, demonstrating an extraordinary mobility. Almost exclusively grazers dependent upon green grass, the apparent randomness may have been related to differing moisture content of the forage and the presence of patches of tsama melons, but above all opportunistic in response to the highly erratic rainfall, oscillating mainly along a south-east to north-west axis. Perhaps in former times attracted to the springs of the Ghanzi area.

There was no evidence to indicate these movements were conducted along regular seasonal or annually recurrent patterns or migration routes, unlike the much smaller Makgadikgadi population which shows a water dependent annual migration pattern. In the wet season this population feeds on the short-grass plains along the west shore of the Ntwetwe Pan, and in the dry season congregates in a limited area of woodland and long grass along the Boteti River.

In the Serengeti, which harbours the only other large-scale wildebeest population remaining in Africa, everything is compressed into one-tenth of the scale; and the rainfall gradient radiates from a point in the south-east with its long axis in a north-west direction. Here there is an annual wildebeest migration roughly from the south-east to the north-west; then funnelled by settlement, rather analogous to the Kalahari movement, to the north-east. From here they move south to the point of origin completing an elliptical movement.

The differences summed are, that in the Serengeti everything is compressed into a much smaller area resulting in an annual migration, or more properly a continuous movement. But although having an overall pattern the direction of movement is variable due to the erratic pattern of rainstorms. The Kalahari migration, on the other hand, results from drought conditions but may be equally variable in direction if there is an indication of possible rainfall in a particular area. Thus southern sectors of the population may move to the south rather than northwards, and in 1979 a large number moved in the direction of the Limpopo River almost 500 km to the east, instead of accompanying the majority of the population to the north. Due to the extensive scale the animals are less concentrated to begin with, and when they move they are in herds of from 40 to 400, scattered across a vast front more than 160 km across (Fig. 6.1).

About 1930 to 1931 a mass die-off of wildebeests was reported from Sebele near Gaborone, north as far as Nata, and east to Francistown. Those dying at Nata may have been from the Makgadikgadi population, which between 1959 and 1965 also crashed. At least 40,000 were estimated to have died in 1959, but this was probably a gross underestimate. In the 1980s drought their numbers fell from an estimated 52,000 in 1979 to 12,000 in 1987, and more deaths followed.

The situation was exacerbated after 1958 by the construction of the completely arbitrarily-sited, as far as the wildlife was concerned, veterinary cordon fence designed to cut the country in two. This was to comply with European Community beef export requirements, that stock raising areas must be isolated from potential foot-and-mouth-disease infected areas; resulting in a far-reaching effect upon the ecology of the northern third of the Central Kgalagadi Game Reserve, concentrating against this impenetrable barrier the herbivores which normally migrated north in bad years.

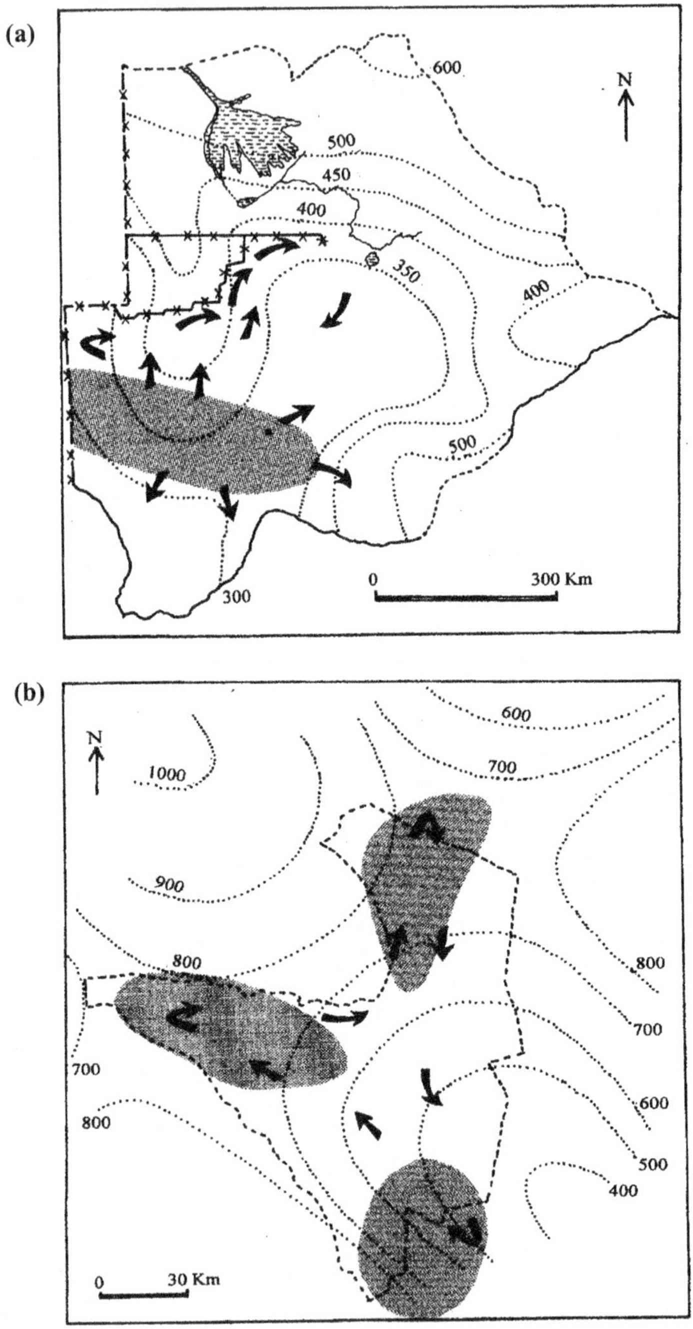

Fig. 6.1 The wildebeest migrations in relation to rainfall for the Kalahari and the Serengeti compared: **a** Botswana **b** The Serengeti. Shaded areas = areas of concentration

Plate 6.2 Zebra numbers collapse suddenly in drought

Although grazing and browsing may have been sufficient, the region being well-wooded, this concentration produced an excessive pressure upon moisture-bearing tubers and bulbs which the herbivores dig up and which, in average years, may suffice to see them through the waterless season. As the plant is killed in the process, over consumption probably led to decline of this important dry season moisture resource, compounding the problem. In 1964, after three years of poor rains, one tenth of the wildebeests were dying in this northern sector every five days; while gemsboks, hartebeests and springboks also declined in numbers.

Zebras which formerly occurred in the northern half of the Reserve and were once abundant on Deception Pan, died out in 1962; although five briefly reappeared on the pan during a wet spell in March 1984. North of the fence the 100,000 estimated to be present in 1979 fell to 34,000 in 1987 (Plate 6.2).

In 1963, 300,000 wildebeest are estimated to have died around Nata, although this seems to be rather a high estimate compared with the total population present in the country in 1979. It suggests they once moved as far eastwards as the Shashe River, although this movement may have related to a separate eastern population. A severe die-off also took place between Lake Xau and Rakops in the 1964 dry season, when at least 15,000 died within four months, the first recorded die-off in this area.

In 1970 another die-off took place in almost exactly the same area, occurring again after a series of dry years culminating in a particularly bad one, although there was still water in Lake Xau. First indications of impending mortality were the massing of wildebeest and hartebeest in the dry season range in greater numbers and earlier in the season than usual. Very large concentrations of wildebeest then began building up in the Lake Xau area in June to July, with many animals dying in September. Conditions continued to deteriorate until mid-November, by which time the wildebeest were driven to the desperate measure of entering villages and eating the thatch from hut roofs.

In 1969 there was also a large-scale movement southwards and records show a significant migration once existed east and south-east from the southern Kalahari to the Molopo River. The earliest reference to this may be that of a few troops which strayed eastwards from the Molopo to within 50 km of Mafeking in 1891, the previous year being recorded as very dry and the Molopo dried up. The first definite reference appears in reports in 1941 by the officer in charge of Werda police station, who in July of that year wrote the area was being "over run by great masses of Wildebeest all of which are moving down to the river looking for water… The Wildebeest are in extremely poor condition and many of them are to be found dead on the Veld, some of them are so poor that they can only just stagger out of the river bank". In October he reported: "most of the Wildebeest seem to have moved off from the river. There are still, however, mobs to be seen and they are in a pityful (*sic*) condition. Many thousands of these animals died, no doubt, from thirst and starvation.

"As a result of the high death rate of game in the B.P. [Bechuanaland Protectorate] Kalahari area, beasts of prey are numerous".

Another report of September in that year commented: "The wildebeest does not possess the drought resisting qualities of the other game of the Kalahari. This whole district is very dry and the wildebeest have been moving south in large numbers in search of water and more succulent grazing.

"Large numbers have reached the Molopo…

"Many dead and dying wildebeest were seen in and around mud holes in the river… Thousands of gorged vultures were encountered along the river. In some cases they could scarcely raise themselves from the ground. One farmer had dragged 32 carcasses out of a muddy water-hole".

The Assistant District Commissioner for Tsabong wrote in January 1942:

"(b) The wildebeest trek South *every* winter [original emphasis] in large herds and many cross over into the Union between Tsabong and Khuis where there is no fence.

"(c) It is only in times of drought that they force their way through the fences and mix with grazing stock.

"(d) Of recent years bad droughts occurred in 1933, 1937 and 1941.

"(e) When the wildebeest are on trek they move southwards steadily and can cover 20 to 25 miles in a day…

"(f) What steps do the Union Government [i.e. South Africa] propose to take to control the large number of wildebeest breeding in the Nosop Game Reserve? These no doubt constitute just as big a menace to the farmers of the Gordonia District as do our wildebeest to the farmers of Vryburg and Kuruman Districts. The Parks Board will probably say that they don't but nevertheless I gather that a good many trekked out last year."

Hundreds of carcases were found along the Molopo River fences in 1980.

6.2 Wildebeest Destruction

Wildebeest are hated by the cattle-owner because they compete for grazing and carry diseases transmissible to cattle. In 1931 the question of wildebeest destruction was raised by a speaker in the European Advisory Council, the forerunner of Botswana's parliament. It was not the first time the issue had been raised, in 1923 African Ranching Farms in the Tuli Block had written to the Resident Commissioner requesting permission to destroy them.

The speaker complained of cattle losses due to malignant catarrh, a virus disease carried by wildebeest which infects cattle from the afterbirth when wildebeest have calved: "There are hundreds of wildebeest. Last season I had considerably more wildebeest on my farm than I had cattle... Last year we were asked to exterminate, and I understood you [Resident Commissioner] were doing this in conjunction with the Governments of Southern Rhodesia and the Northern Transvaal. As regards extermination of Protectorate wildebeeste in my district, I may say that not half the increase was destroyed. The Chartered Company agreed to the extermination, and on the British South Africa Company's ground 300 head were destroyed".

In October 1946 the issue was again debated, the problem area being the Tuli Block farms on the Limpopo River. The wildebeest were not reported as migratory, but they may have been animals which stayed behind after migrating to there in the 1941 drought: "The question of the destruction of wildebeeste has cropped up several times in the last three or four years. Many people are complaining. During the war years ammunition was very scarce and all game increased tremendously. Many farms which had been abandoned for some years have become absolute game reserves. Some farms which were never taken up before and which were never shot over much are also full of game and the particular species that is constituting a real nuisance is wildebeest...

"About twenty years ago [a reference to the letter the Resident Commissioner had received from African Ranching Farms] the late Mr. H.C. Weatherilt complained very bitterly in this Council of the same thing and asked for a free hand to destroy wildebeest; we were allowed to shoot twenty-five, which is useless when the wildebeest are in hundreds coming and going as they like".

The speaker continued that it was impossible to carry out progressive farming where wildebeest abounded and that they should be taken from the list of protected game and treated as vermin, shot on sight without licence or permit.

But it was not until 1953 that the landowners got their way and the wildebeest was classified as "Small Game" throughout Botswana, a legal manoeuvre which by downgrading it from "Large Game" meant that it could be hunted without a licence. This had already been applied since 1924 from time to time for limited periods and in specified areas to help control migrating wildebeests. The classification was revoked in 1961 when the wildebeest became a large game animal again, huntable only on licence.

The legal manoeuvre had been applied in the Tuli Block in July 1924, July 1928, January 1930 and October 1935; suggesting wildebeests migrated in these years as far east as the Limpopo. Whereas the change usually ran for one year, the order of 1935 was not revoked until September 1938. African Ranches, for example, complained in 1935 that 600 wildebeest had been shot on the farm and 2000 still remained, while cattle had to be sold because of lack of grazing. Hunting wildebeest did not pay unless one could shoot 40 per week, one hunter achieving 82 in 14 days. The method was to follow the herds, shooting as many as one could. A hunter on African Ranches manufactured 900 kg of biltong, making 225 kg per trip, or 135 to 225 kg per month.

In the Francistown District the measure was applied from 22nd January to October 14th 1930, thus identifying the time when the migration eastwards took place. In 1939 it was applied to wildebeests in the Bakgatla Reserve, and in 1941 to part of Kgalgadi District with an amendment in 1942. These latter orders coincided with a massive southerly migration across the Molopo River into South Africa in 1941. Similar migrations were reported for 1927, 1935 and 1937. In 1942 the Notice for the Bakgatla Reserve was renewed. It was also instituted in the Tati (Francistown) District.

But wildebeests and other game were being savagely destroyed by biltong hunters from South Africa in a return to the excesses of a century before, as one member of the European Advisory Council revealed: "The destruction of game last season [1945] was a disgrace; while the Union Government is proclaiming more and more Game Reserves, the Union biltong hunter does his shooting in the Bechuanaland Protectorate. A man is allowed to take out of the country 20 lbs of biltong free and pays 6d. per lb for all above that amount. Parties of up to eight men each visit the Protectorate in vans, 3ton trucks and motor cars. One or two members of the party apply for a large game licence, the rest of the party sight-seeing, but carrying arms in case they meet lion. Permission is obtained to shoot four large game-there is no restriction on the small game including Springbok... When returning they camp just outside the Molepolole village, passing through during the early hours of the morning... Often a three ton truck passes during the night fully loaded, while the next morning another member of the same party, travelling by car or light truck, does the reporting.

"One Springbok will produce approximately 7lbs dried biltong, Hartebeeste probably 20lbs or less.

"From experience, I found that parties out shooting wound at least 4 buck to each buck actually bagged... 1544 lbs of biltong was declared and paid for in 17 payments at Molepolole. 17 people declared having over 20lbs weight of biltong. This was for export or taken out of the country by hunters from the Union. 15 small game licences and 51 large game licences were taken out by non-residents. 17 paid for biltong, what did the others do? Biltong is worth anything up to 10/6d. per lb. in the Union.

"It is alleged that 20 odd troop carrier lorries were out shooting. This does not include cars and light vans...

"I was told of one man holding the record in his party by alone shooting 85 Springbok. This means another 255 wounded and lost... One Transvaal hunter obtained a permit to export 600 lbs of biltong. It was rumoured he had over 800 lbs. 800 lbs means 100 Springbok at least killed and 300 wounded and dead in the veld

(400 in all). I personally saw troop carrier trucks loaded with bags of dried biltong as though it were grain."

He reported the same thing apparently took place in the Bangwaketse Reserve and therefore probably "over the whole of the Protectorate." However, another Member pointed out that when "open shooting" was allowed in Ngamiland the Chief appointed a native to accompany each shooting party to see they remained within a certain area. Likewise before the District Officer could issue a licence in the Bamangwato Reserve, he had to have the written consent of Chief Tshekedi, the Chief usually selecting the party whom he wished to give a permit to, and no licence was supposed to be issued to shoot in any native reserve unless the permission of the Chief had first been obtained.

The Acting Deputy Resident Commissioner stated: "I do know that some Chiefs have been very much against hunting parties going into their reserves. Chief Bathoen is one of them and Government itself has become increasingly alarmed at the amount of shooting going on. The game did have a respite during the war years because ammunition was practically unobtainable, but we have, in this last open season, been inundated with applications… Actually we would like to stop shooting altogether but we feel that it is not possible to make it as rigid as all that since there are times when bona fide sportsmen require permission and it is not desired to close the door to everybody. These are the views of government which have been sent to District Commissioners; they have been asked for suggestions as to the procedure to be adopted next year in this matter. We are very much concerned about this indiscriminate shooting of game".

Another speaker pointed out that in the Bakwena Reserve the Chief's permission to hunt had to be obtained, but suggested with the present high price of biltong one could expect bribes to be offered and: "A party has the right to be out for say three weeks. During the first week during darkness the lorry will leave the country with its biltong and return. This will probably take place several times. It is difficult for a Chief or for the local Police to check up. I think the Chiefs too appreciate that the slaughtering of game should be checked."

A motion, that "timely steps be taken to end the wholesale slaughter of game in the Southern Protectorate", was then put to the vote and carried unanimously. In fact the hunting of large game on Crown land north of the Molopo River had been forbidden since 1924, the area later being specified as Ghanzi and Kgalagadi Districts; but it was not until 1961 that comprehensive new legislation was introduced.

6.3 Hartebeest Decline

In the 1930s it was alleged the hartebeest found in Botswana was the Cape or red hartebeest, extinct in South Africa where it had once been widespread. Others believed the Botswana animal to be a race known as Selborne's found in Namibia, and in South Africa only in captivity on de Beers's Kimberley Game Farm. After careful study, Smithers, of the then National Museums of Rhodesia, came to the conclusion in 1971 that Selborne's hartebeest was nothing more than the Cape hartebeest and did not even warrant the status of a distinct race. Specimens from Botswana conformed closely to the descriptions given by Sparrman in 1784 and Cuvier in 1804 of the original Cape hartebeest.

Outside of South Africa the hartebeest had not become extinct at all but had survived in surprisingly large numbers. Although subjective estimates put the number dying against Ghanzi farm fences in the 1981–1987 drought as "10,000", probably an exaggeration, its 1980s decline did not receive as much attention as that of the wildebeest. Aerial counts of hartebeest conducted in 1986 to 1990 suggested a significant decline for this species also, in excess of 70%. By 1990 there were less than 36,000 in the country compared with about 300,000 in 1979, their numbers now slightly exceeding those of the wildebeest. Although they do not appear to have migrated in such numbers as the wildebeest, being a more drought tolerant species, their numbers are highest in the northern Kalahari in the dry season, suggesting a somewhat similar movement pattern.

A speaker in the Legislative Council in 1961 referred to hartebeest in 1959 and 1960 migrating "in their thousands" just on the edge of the Bangwaketse District. The same speaker noted that in 1921 to 1922, from Sekoma to Kgokong and Dutlwe, you could travel for 65–100 km "and never be out of view of game migrating" trekking south-west or north-east. Although the species are not mentioned they would have been wildebeests, hartebeests, springboks and perhaps, elands.

Selous stated that the hartebeest had the same range as the gemsbok and was not found north of the Makgadikgadi Salt Pans, being unknown in the country between the Chobe and Mababe rivers.

6.4 The Mystery of the Springbok

The year prior to Chapman, Galton somewhere between the western border and Gobabis in Namibia, noted that he passed a great herd of springboks migrating in September, adding the significant remark: "It was by no means as numerous a herd as is often seen in Bechuana country", but the tufts of white hair on the backs of the bucks were as thickly scattered over the country "as daisies on a lawn." In 1877, returning from Zimbabwe, Barber stated that on the Nata flats in Botswana he came across vast herds of wildebeest and springbok, one morning passing for three hours through a mass of game without a break anywhere, the springbok being on trek.

In South Africa in 1849, just before the Orange River, Livingstone saw the last portion of a migration of springboks which he noted came from the great Kalahari Desert and, "when first seen after crossing the colonial boundary, are said often to exceed forty thousand in number. I cannot give an estimate of their numbers, for they appear spread over a vast expanse of country… as they come from the north about the time when the grass most abounds, it cannot be want of food that prompts the movement. Nor is it want of water, for this antelope is one of the most abstemious in that respect." He considered they sought short grass areas where they could see the approach of a predator and "The Bakalahari take advantage of this feeling, and burn off large patches of grass, not only to attract the game by the new crop when it comes up, but also to form bare spots for the springbuck to range over.

"It is questionable whether they ever return, as they have never been seen returning as a body. Many perish from want of food, the country to which they have migrated [south of the Orange River] being unable to support them; the rest become scattered over the colony".

Livingstone concluded, "It is probable that, notwithstanding the continual destruction by firearms, they will continue long to hold their place."

In the mid-1870s Selous recorded the springbok as common in many parts on the borders of the Kalahari desert, and on the Makgadikgadi and Ntwetwe salt pans, but its northern range was bounded by the forests running east and west south of the Mababe river. Between Mopipi and Matima Ford on the Boteti, Nicolls in July 1889 "every day noticed large troops of wildebeest, Burchell's zebra, and thousands of springbok." In the 1920s large flocks still occurred in the Nata area but the species seems to have been finally shot out there in the early 1980s.

Of all the antelopes it is the springbok which is best known for its mass migrations, or "irruptions". In 1896 in the western region south of the Molopo River, a greater "trek-bokken" than had been seen for many years took place. In the nineteenth century two significant treks southwards across the Molopo River out of Botswana occurred, one in 1946, the other in 1950. The former took place in October of that year and was over ten km deep, a solid mass representing many thousands of animals, which moved at walking pace down the Nosop and into the South African farming areas. Some reached Upington, more than 360 km from where they were first seen. The second movement began from southern Botswana in the Tsabong area, the animals becoming restless and massing together, finally moving in November, crossing the Molopo River on a 320 km-wide front into the northern Cape Province area. Populations of the central Kalahari were probably the origins of the mass movements recorded in South Africa in the nineteenth century. The last occurred there in 1959.

Subsequent to the last great trek of 1896 they have never reached the same numerical proportions, but this is probably not due so much to recurrent drought mortality itself, as to the "step-down" effect, whereby, after major mortality, as may have occurred with the rinderpest panzootic in 1899 (but no mass die-offs of springbok are on record), the population does not have time to build up again to its former level before another recurrent drought comes along, even if there was the land still available for it to colonize. In the'fifties it was considered springbok migrations still took place about every five years, but from 1945 or 1950 there was a very large decrease in the numbers of animals; although there is one report of large herds being seen in November 1952 between Tsabong and Ghanzi.

In 1990 their numbers throughout the country were estimated at just over 117,000, but could have been as high as over 200,000. A tiny remnant compared with the astonishing half a million that Cronwright-Schreiner estimated to be in view at once south of Kimberley in the north of Cape Province in 1896. He estimated a swathe of country over 160 km long by 24 km wide was covered by a solid, moving, unbroken mass (Plate 6.3).

Plate 6.3 Springboks on the Nosop river

6.5 Eland-Large Troops Still Wander

Livingstone was the first to record herbivores which did not drink must rely on preformed water in their food. He noted that although water was inaccessible to them, large numbers of elands were found in good condition and their stomachs contained considerable quantities of water. Nevertheless he correctly believed they could subsist only where there was some moisture in the vegetation on which they fed: "for in one year of unusual drought [1851] we saw herds of elands and flocks of ostriches crowding to the Zouga [Savuti?] from the Desert… As long as there is any sap in the pasturage they seldom need water." In his private journal for September 1851 he wrote "Immense herds of eilands [*sic*] have been seen all along our route [travelling from Lake Ngami to west of Serowe]. They have been compelled to come out of the desert by the drought being unusually great, and all were very lean. Many had mange or scab." He noted "The drought must be intense in the desert when these [ostriches] are compelled to come to the river for water." Great numbers were caught in pitfalls near the river.

In 1910 Vaughan-Kirby wrote that he was struck by the comparatively low condition of "all" the eland he met with on a trip with Hodson and the Resident Commissioner to northern Botswana: "Of course these animals are never in really prime condition between mid-October and mid-December; but this was only the end of September, yet they were all decidedly "poor-that is, for eland".

In the mid-1870s Selous recorded the eland as plentiful in the Kalahari desert to the west of the Bakwena and Bamangwato, but they never now came as far east as the road between Molepolole and Shoshong. North of Shoshong along the roads leading to Lake Ngami and up to Tati there were always a few to be found but they were usually very scarce; "at times, however, large herds wander out of the Kalahari desert, as far eastwards as the waggon-road [leading from Shoshong to Tati]... These migrations usually happen in the months of February and March, and the Bushmen say that they are to be accounted for by the fact that at that time of year a particular sort of small bush is in leaf, of which elands are very fond." Later he wrote it was the berries of the bush which they sought and the largest herd he saw was one over two hundred strong. In all the land between the Boteti and Chobe they were still to be found in some numbers, often herds of over 100 together were seen and they were particularly plentiful in the dry desert country through which the Chobe flows.

Along the Chobe he noted the elands drank regularly at the river and then fed up to some ten km or more away from it, but in other regions they were able to thrive without water and he considered that the desert elands attained a greater size than those found in the well-watered parts.

Bryden referred to the eland formerly occurring in southern Africa in troops of up to 100: "Even at the present time [1890] in the North Kalahari solitudes, large troops still wander. I have seen great quantities of spoor to the right of the road between Inkouane and Kanne; and in the same region a good troop seen... contained between thirty and thirty-five individuals".

Eland also sometimes migrated southwards in large herds, as in 1985 when they crossed the Molopo into South Africa, the farmers breaking down the fences to let them enter so they could then slaughter them. In 1990 it was estimated there were 26,500 eland in Botswana, although it could have been as many as 40,000. Of these the number in the Central Kgalagadi Game Reserve does not seem to have changed much since 1979, with estimates of almost 7000 then and 7500 in 1989.

6.6 Gemsbok

Characteristic of the Kalahari desert, but the most widely dispersed species in the country, is the gemsbok, although its numbers tend to be less than those of the other large antelopes and it does not move in big herds. In 1979 there were something like half the number there were of wildebeests and hartebeests, but being more drought tolerant in 1990 they outnumbered these latter species by a factor of three or more, with an estimated total for the whole country of almost 114,000. Oswell recorded they were scarce, and hardly met with except in the barren open stretches of country like the Kalahari desert, and there were more near Cape Colony than further in: "He can do without water for a long time certainly-indeed I believe altogether." Oswell noted it ate certain roots such as the large tuber *luhoshe* and the tsama melon for its water requirements.

In 1871 Thomas Baines came across three at the Lotsane River near Palapye: "which I have often seen in western or central [southern] Africa but did not expect to meet so far east, as it generally inhabits only the arid plains of the interior. It was once common as far as Sechomo's [Sekgoma's] (now Matjen's) [Macheng's] and

Plate 6.4 The gemsbok, a water independent species of the Kalahari

probably the drought at present prevailing has tempted these so far. We saw seven or eight more".

Selous noted it was plentiful in the Kalahari desert and occasionally met with on the route to Shoshong, becoming abundant if one penetrated westward but unknown to the east of the route. Between Shoshong and Tati there were a few around Palapye, Serowe and Shashe; and sometimes they were found on the upper reaches of the Motloutse, Shashe and Tati rivers, some even wandering as far eastwards as the Ramokgwebana river. Others were found near Tamasanka and Damasetshe pans on the eastern border. They were numerous around the Makgadikgadi Salt Pans. In his day he considered it was never found between Pandamatenga and the Mababe river, although it was recorded there by Hodson in 1906.

Even gemsbok appear to have declined by almost a third in the great drought of the 1980s. Before this it was alleged they were declining due to excessive destruction by hunters with vehicles, and San on horseback. They are more vulnerable in some respects than other species as they do not move far from one place. Even in severe droughts they have no need to displace great distances to satisfy their requirements (Plate 6.4).

6.7 The Red Lechwe

Another antelope formerly occurring in vast numbers, but of a quite different habitat, is the water-loving red lechwe, a relatively rare species enzootic to the wetlands of south-central Africa. This animal was once numerous on the Boteti River where it is now extinct, as far downstream as Lake Xau. When in 1849 Livingstone returned along the river after reaching Lake Ngami, he reported encountering "an entirely new species of antelope called leche or lechwi", east of its confluence with the Thamalakane River. In a letter dated January 10th 1850 from Cape Town to his hunting friend Captain Vardon, who had returned to England, Oswell recounted the journey of discovery to the Lake. He wrote that eight or nine lechwe were shot, the first by Piet, his wagon driver, the second by Livingstone, and the third by Mungo Murray, Oswell's hunting companion. Possibly Oswell sent Vardon the skin and horns of a lechwe with this letter for Vardon presented these to the British Museum, the first such specimens of the antelope to be recorded.

In 1862 Chapman saw herds of 50–100 near Kwaraga, and hundreds were reported in 1865 where the Thaoge River entered Lake Ngami.

At Lake Ngami, as on the Boteti, it probably became extinct about the end of the nineteenth century. It was reported as scarce there by 1892. On the Boteti near Mopipi in July 1889 Nicolls found it very shy due to continual persecution by the locals for its skin. The reed beds south of Rakops have now virtually disappeared, so habitat decline as much as hunting may have caused its extinction in this area, as at Lake Ngami also. The nearest now to this former range are found west of Maun in the Okavango swamps, where, divided into small groups, more than 30,000 are estimated to still exist.

On the north bank of the Chobe, before its confluence with the Zambezi, Livingstone reported lechwes in prodigious herds, feeding quite heedlessly all over the flats, although the numbers of them and of the sitatunga which were killed annually he thought must be enormous. In his private journal he wrote "On awakening this morning Lechwees were feeding close to the waggon and village in perfect confidence. It is a lovely sight to see the game so tame, but fire arms will soon dispel all this."

In the 1880s up to 700 lechwes at a time were killed in local hunts on the north bank of the Linyanti using spears and dogs. Now this population numbers only around 1200.

Describing the swamps in the Kachikau area in 1874, Selous wrote that herds of lechwes were to be seen standing about in the shallow water. Travelling down along the mainland south of Kataba he saw, out in the marsh, "numberless lechwes". He wrote this antelope was first met with in the marshes of the Botete river, and was very numerous on the grassy plains adjacent to the Thamalakane, Mababe, Mochaba, Savuti and Chobe rivers.

At the end of the 1970s some 15,000 were estimated to occur in the Linyanti swamps, much of which area lies within the Caprivi Strip, several thousand of the lechwe being along the edges on the Botswana side. In 1962 more than 2500 existed along the Chobe waterfront, but these had fallen to less than 800 in 1966, 550 in 1971 and 162 in 1990; the drying-out of the swampy areas making them more accessible to poachers on foot (Plate 6.5).

Plate 6.5 Red lechwe rams. Once in thousands in the Chobe area, they have now almost disappeared there

6.8 The Zebra

The zebra formerly occurring in southern Botswana was probably the first species to become extinct in the country over a large part of its range. Today it is not normally encountered south of latitude 22°, but once could be found down to the Molopo River, contiguous with the herds in South Africa Cornwallis Harris in 1837 had described as occurring in "incredible numbers". Bain reported an enormous herd at Korwe in 1826 (25°25′S) and shot several near Kanye. In 1834 he captured six near Phitshane on the Molopo. Oswell encountered large herds at the Molopo in 1845 and in the same year Cumming recorded zebra in the "Valley of Bakgatla" and immense herds about Shoshong in 1847. Anderson reported them in the region of Mochudi and on the road to Shoshong in 1869, while Barber came upon a great herd north of Serowe in 1875. In 1879 Parker Gillmore reported the Mhalatswe valley, 200 km south-east of Shoshong, (in fact the distance was not more than 75 km), as 'fairly swarming with springbok, veldebeest [*sic*] and Burchell's zebras'; but by 1892 Nicolls and Eglington were to write: "this Zebra is now rarely seen in a direction South of the Botletle River, although after heavy rains it is not uncommon to find a few straying farther into the Kalahari."

The zebra which had vanished over such a vast part of the range in 50 years was the true Burchell's zebra, *Equus burchelli burchelli*, the race first collected by William Burchell some 200 km south of the Molopo River in 1822–23.

Those surviving in the north were thought to be of two races: Chapman's zebra *E. b. chapmanni*, a race of the Zambezian Region; and the Damara zebra *E. b. antiquorum* from Angola, an extension of the Zambezian Region. The races are distinguished by the latter never having the legs wholly striped while the Chapman's has legs striped to the hooves. In the north-east these two races are believed to have intergraded producing animals with very variable leg patterns.

Zebra in the eastern Ngwato region which became extinct in the 1960s drought, and those in the Tuli Block which succumbed in the drought of the early 1980s, were of the northern races.

That a race which existed in incredible numbers disappeared so rapidly throughout its range, was probably attributable more to increasing aridity of the area than to the systematic slaughter for its skin, which undoubtedly took place, although there are no records of the latter, such reports as there are referring to "antelope skins". Requiring water and good grazing, the zebra's decline seems to have been already well advanced by the middle of the nineteenth century and hunting probably had little effect upon its downward population trend. Livingstone reported great numbers dying from symptoms similar to "horse-sickness" (equine piroplasmosis, a tick borne blood parasite caused by *Babesia caballi* and *Theileria equi*)), a disease more likely to have been anthrax as zebra are considered very resistant to horse-sickness; but perhaps an indication they were already succumbing to environmental stress. Either way, the Burchell's zebra in Botswana illustrates just how rapidly a race, and therefore a species, although formerly existing in large numbers, can disappear completely from an area.

In the 1980s drought, zebra in the region north of Nata and the Makgadikgadi Pans, showed not a gradual decline like other species, but an abrupt population collapse to a low level. Just as they suddenly disappeared from the northern Kgalagadi Game Reserve.

In the Serengeti the zebra has been found to facilitate wildebeest grazing by feeding on the coarser grasses, thus exposing the softer, what effect their disappearance in parts of Botswana may have had on wildebeest numbers is not known.

6.9 A Fragile Resource

The southern zebra apart, which of these once numerous remaining animals, one might ask, will become extinct first? The lechwe, the wildebeest, or perhaps even the springbok? Will it be the story of the passenger pigeon in America all over again? The story of numbers being so great it is inconceivable they could ever become so low as to vanish into oblivion? It took only a few years for the market hunters to reduce the population of passenger pigeons to a point at which recovery became impossible. It may be a feature of the biology of species which live at high densities that they are unable to persist at low numbers for social or genetic reasons. Certainly they seem to die out rapidly when low thresholds are reached.

As records reveal, the decline of the Kalahari wildebeest has been engendered not solely by its nomadic habits in time of drought being interrupted by cordon fences and through competition for grazing resources, but also as a direct result of its conflict with stock farming, and decline appears to have been already well advanced by the 1960s. Such a process is considered inevitable in the face of modern agricultural development policies. The Kalahari wildebeest is a prime example of a species once considered so abundant as to be regarded as vermin, of which the constant attrition in numbers has resulted in sudden near-extinction in parts of its range, or at least the altering of its social dynamics from one of macro gregariousness to one of sparsity.

Where permanent water exists the zebra is a successful competitor with the wildebeest. Thus in the Chobe-Linyanti dependant ecosystem they outnumbered the wildebeest by 4.5:1, with a total population at on etime of 63,000. In the Makgadikgadi ecosystem the ratio was almost 2:1, with a population of 100,300 until the deaths in the 1980s. Had the zebra not already disappeared in southern Botswana by the twentieth century, it too would undoubtedly have been regarded as vermin.

Theoretical analysis shows that most species populations characteristic of the Kalahari: the wildebeest, hartebeest, springbok, gemsbok, kudu, eland and ostrich; could withstand high levels of mortality without leading to further population decline relative to that experienced in 1986. However the springbok, having shown the least decline, can withstand the least mortality. No population can recover its 1979 level in the southern Kalahari without passing through another drought cycle and having its numbers knocked down further, on the assumption that such a cycle will take place within 20 years. If harvests were three times the given official hunting quotas in the 1980s, not an unrealistic assumption when predation, illegal hunting and other unrecorded harvests are taken into account; then gemsbok, kudu and ostrich, all regarded as common species, would eventually become extinct. The kudu in as short a span as 20 years, with greater than 95% of the population lost in the first eight. For ostrich and gemsbok 95% of the losses could occur in 56 and 73 years respectively.

This suggests the large herbivore resource of the Kalahari, is, under a regime of recurrent drought, a fragile resource. It is not the superabundant resource with a considerable potential for exploitation that has been suggested. It is hazardous to say "use it or lose it". Analyses indicate the majority of the principal herbivore species require many years to recover their numbers following Botswana's drought which ended in 1987. Some can only effect a partial recovery, while the eland may apparently never do so, but gradually decline to extinction in this habitat. In the case of the zebra, more water-dependant than these other species, as we have seen it was not a gradual decline but an abrupt population collapse.

In the 1960 to 1965 drought, zebras and giraffes disappeared from the eastern Ngwato, giraffes from the Ngwaketse and much of the Kweneng, tsessebes and reedbucks from Nata, waterbucks from the Notwane, and sable antelopes from the Tati farms. They disappeared due to excessive hunting, hunger, and thirst.

It has been argued the mass mortalities of wildebeests were caused by starvation, either because there was no water in the dry hay and therefore they could not eat it, or in the proximity of water overgrazing by cattle had removed all the available forage. But had there been sufficient moisture in the food they would not have had to look for free-standing water. Thus water, rather than food, is apparently the limiting factor in the Kalahari, and it would suggest the majority of drought-induced mortality is largely density independent at the peak stocking levels which have been reported, which have adequate food but not water.

Thus far from representing a wasted, superabundant resource in times of good years, high population build-ups may ensure an adequate number of survivors after a drought, for the number dying will not be related to the number initially present. In other words, if drought kills off 90% of a population, it will do this whether the population numbers 300,000 or 30,000. The dangers of culling are therefore obvious. The Kalahari is not a cornucopia.

6.10 The Tragedy Continues

At the end of a drought in 2011, a survey of the Ngamilamd District to the west of Chobe National Park and encompassing the Okavango Delta, revealed that in the fourteen years since 1996 drastic declines had taken place in species. Ostrich numbers had apparently fallen by 95% from an estimated 11,900 to 500; 90% of wildebeest; 84% of tsessebee; 81% of kudu; some 60% of giraffes, and lechwe numbers by 50%; but elephant numbers held up. These declines were attributed to a combination of factors, not only drought but poaching and land use changes. At the end of the year came torrential rains and flooding, and in 2018 it was estimated there were almost 89,000 lechwe, 60,000 zebra, 16,485 wildebeest and 12,660 hippopotamus; but the following year drought, out of phase with a 20-year cycle, returned, killing large numbers of hippopotamus and elephants.

Chapter 7
Disease and Death

Abstract 1942 tsetse report promotes destruction of game. Cattle numbers. The 1933 Pim report on economy lays basis for development of cattle industry. Animal diseases. Early reports. The great nineteenth century rinderpest plague. Effect on cattle and game, large numbers of game survived. Foot and mouth disease cause of veterinary fences to allow beef exports. Buffalo its carriers. Large buffalo populations in north. Calls for their destruction. Tsetse fly, sleeping sickness, and cattle trypamosomiasis. Nineteenth century reports of tsetse fly. Fly linked to buffalo. Decline in tsetse fly. Attempts to eliminate by destroying game.

Keywords Big game · Botswana · Diseases · History · Quotations

7.1 The Pim Report

At the beginning of 1940, Dr. J. Macaulay arrived in Maun. He was seconded from Kenya to conduct a survey of the area infested with tsetse and trypanosomiasis. In his report, published in 1942, he calculated 45,272 head of cattle worth £96,000 were at risk, and came to the conclusion that because of the sparse human population and low economic level of Ngamiland, there should be a long-term policy of tsetse control. He voiced an implacable hatred towards game. Game was a luxury which could not be afforded, he stated, and recommended its organized destruction and the encouragement of hunting. Furthermore, game reserves should be discouraged, water points outside of the Okavango Delta poisoned, and the deadly disease of rinderpest re-introduced (with the cattle of course protected from it). As a result of his recommendations of butchery a game destruction programme was instituted in the Maun area.

In 1961 the then Director of Veterinary Services stated he had been opposed to the creation of the Chobe Game Reserve (now the Chobe National Park) from the start: "our cattle industry can never be put onto a proper footing if we are going to consider the game." The following year the Nosop Reserve (now the Kalahari Gemsbok National Park) came under attack. It was a liability and best handed over

C. Spinage, *Big Game of Botswana*, https://doi.org/10.1007/978-3-031-60033-3_7

to South Africa. He stated it could not be guaranteed that it would not be de-gazetted at some future date.

Cattle had co-existed with game in Botswana for centuries, and this bigoted hostility stemmed principally from the fear engendered at the end of nineteenth century by the rinderpest plague, which swept the length and breadth of Africa causing incalculable losses.

While there are few early records of cattle numbers, Bain estimated in 1826 the Ngwaketse chief Sebego seized at least 2000 head from the Mma-Ntatisi at Dithubaruba, which he claimed were his cattle that had been stolen. In the 1870s there were over 200 cattle posts in Khama's territory, many of them along the Mahalapye River, in addition to posts for sheep and goats. In 1895 the total numbers of cattle were estimated at over one million. By 1904, due to rinderpest they had fallen to an estimated 139,000.

But numbers soon recovered, and beginning in 1920, by 1925, 10,000 cattle were swum annually across the Chobe River at Kazungula after having spent several months grazing along the Chobe waterfront to fatten them. These cattle came from Ngamiland, the Tawana owning about 154,000. From 1921 to 1932 an average of 29,000 head a year were officially exported, of which 93% were native-owned. By 1930 the total number of cattle in the country was estimated to be back at one million again and it was just under that in 1949. In 1959 it had risen to over 1.3 million, and in 1966, after the drought, it stood at 1.2 million. But then rising to three million in 1981 it fell back again to 2.7 million in 1984, with densities in the grazing areas at least double the recommended stocking rate. In 2011 the number was still 2.7 million, which fell to 1.7 million in 2015 and 1.5 million in 2021. In addition, sheep and goats, which remained largely unaffected by droughts and disease, rose steadily in numbers to one million in 1984, of which 90% were goats. In 2022 the numbers had doubled to 265,000 sheep and 1.7 million goats.

Rey, the enthusiastic Resident Commissioner of the Bechuanaland Protectorate from 1930 to 1937, was anxious to develop the economy of the country, especially through cattle exports. His ideas were greatly supported by an economic review of the country conducted by Sir Alan Pim, whose report in 1933 laid the basis for the development of the cattle industry.

Apart from a few years, since 1898 the economy had been kept fairly well balanced, revenues for most years exceeding income. But in 1929 expenditure began to rise and in 1931–32 and 1932–33 exceeded income by a greater amount than it had ever done before, with a deficit averaging 50%. Only cattle were seen as providing the answer in this near-desert country; although Rey considered wildlife could also provide important revenue and especially favoured the development of a Chobe national park and a hotel at Kasane. But henceforth all efforts were to be directed towards developing the cattle industry, a tradition assiduously followed to this day.

The first setback was in January of that year when there was a serious outbreak of foot-and-mouth disease. It had erupted in Zimbabwe [Southern Rhodesia] the year before, passing to Botswana through Francistown, and halted all normal dairy production and development, practically closing the country to all external markets

Plate 7.1 The deadly Kuke fence, severing the heart of the Kalahari

for its animal and vegetable products for almost a year. Outbreaks in 1949–50 and 1957–58 resulted in the construction of the Kuke fence in 1958, its 286 km reaching halfway across the northern two-thirds of Botswana. The fence which changed the face of the Kalahari's great game herds (Plate 7.1).

7.2 Animal Diseases

From inquiries among the San Bushmen and Bakgalagadi, Livingstone learnt that many diseases prevailed among the wild animals. He reported seeing wildebeests, hartebeests, tsessebes, oryxes and giraffes, so mangy as to be uneatable by the natives. In the 1950s large numbers of wildebeests were said to die annually from 'mange', a hair loss condition caused by mites perhaps reaching epizootic proportions when the host populations are at high levels.

The first recorded epizootic disease to affect cattle was lung-sickness, or contagious bovine pleuropneumonia, a disease which does not seem to affect game. Prevalent in Europe, lung-sickness was introduced into South Africa in 1854 by a Friesland bull imported from Holland and landed at Mossel Bay, and rapidly carried in all directions by trek-oxen. In September 1857 Robert Moffat recorded the disease had not reached the other side of the Ramokgwebana river, but in September 1859 it was at Shoshong, probably introduced by Robert Moffat's son and another missionary who had bought infected cattle from the Bakwena. When Robert Moffat reached the Goshakwe river in Zimbabwe in October of that year, he sent a message to Moselekatse to send oxen to meet him as he feared introducing the disease if he used his own, one of which was sick. Moselekatse's reply was for Moffat to proceed and if the epizootic should take effect he would be guiltless. Moffat however convinced him of the danger to the vast Matabele herds, most of which were plundered cattle, and the Chief sent people to meet him and bring on his waggons without the oxen, his men pulling them by hand. In spite of Moffat's caution, the disease was brought in from the south in 1861, and by April 1862 Moselekatse's cattle were dying of it. Many thousands succumbed, mortality in some kraals being from 50 to 59%.

In 1879 lung-sickness was extremely prevalent round Kanye where eight to ten head of cattle were dying each day, but by November of 1884 it had not reached Lake Ngami where there were enormous herds.

Writing of the Zambezi valley in Zambia around Sesheke in 1853, Livingstone noted the Makololo preferred it only as a cattle-station although the herds were frequently thinned by an epizootic disease h he attributed to "peripneumonia", but the cattle bred so fast the losses were soon made good. This disease could apparently not have been lung-sickness for as far as is known it had yet to be introduced.

In 1871, an epizootic of red-water fever, or babesiosis *Babesia bigemina*, a fatal tick-borne disease occurring in moister areas, broke out in Natal and by 1874 was making its appearance on the borders of Zimbabwe, carried by trek-oxen. Mindful of his predecessor's misfortune, Lobengula imposed a *cordon sanitaire* forbidding trek-oxen to be brought across the border, and by 1876 had installed a detachment of Matabele soldiers at Tati to enforce the ruling.

Livingstone reported great numbers of zebras found dead in 1845–7 with masses of foam at the nostrils as occurs in "horse-sickness", but this was probably the bacillary disease anthrax *Bacillus anthracis*, the first record of it in southern Africa. Although prevalent in southern Africa and reported as early as 1719 but only affecting horses, there was no recorded outbreak of African horse-sickness in those years, the last having been in 1839 followed in 1854–5 by what is regarded as the most virulent on record. The zebra is regarded as highly resistant to the disease. The bacterial disease anthrax never made a clean sweep of the whole cattle of a village. People who ate the meat of diseased animals could develop a carbuncle and die (although it was probable the infection was caused by handling the meat). It also attacked wild animals, and just over the border from Lobatse in 1845 he recorded 25 kudu dying from it, and that great numbers of wildebeests and zebras perished from the same cause, "but the mortality produced no sensible diminution in the numbers of game".

7.3 The Great Rinderpest Plague

Concern at the persistent spread of lung-sickness was soon to be overshadowed by a far more virulent epizootic-rinderpest. This highly lethal viral disease had ravaged the continent of Europe from earliest times and had a mortality rate of over 90% in cattle. What became a continent-wide panzootic was first recorded in Ethiopia in 1889, probably introduced by infected cattle via the Sudan, outbreaks having occurred on the lower Nile with infected cattle imported from Jeddah on the Red Sea.

Affecting most hoofed mammals to some degree or other, of the wild game buffaloes were the most affected, followed by elands. Not only did it virtually wipe out all of the cattle in its remorseless passage, but untold populations of game as well, although it may not have affected the latter as disastrously as was generally supposed, judging by subsequent rates of recovery.

Following four years of drought in southern Africa, in February 1896 it had already crossed the Zambezi on its passage from north to south through the Continent. Game was dying beyond the radius of a few km around Kazungula but it was not yet affecting cattle there, although all of the 120 odd mission cattle at Sesheke, 65 km to the north-west of Kazungula, were dead and to the west of Kazungula it had crossed the Chobe river and was travelling south. It reached Tati about the middle of March, but apparently had not affected the route, a narrow strip 650 km long, perhaps because of the scarcity of game along it, long shot out.

Captain Gibbons was at this time trying to reach the south from Zambia before his oxen were affected, and before reaching Tati, near the Monarch mine in Zimbabwe, he recorded "This was the first occasion on which the oxen had been right into the infected area, where cattle were dead and dying everywhere." On the second day after leaving Tati his attention was attracted to a yearling kudu standing behind a bush "on the very edge of the track. The oxen were pulled up alongside the poor brute, which was soon seen to be suffering from rinderpest, nor did the animal go away till startled by the crack of the whip as the oxen were once more started, but had allowed me in the meantime to examine him from a distance of six feet." Little wonder the plague spread south.

To enter Palapye with oxen one had to first obtain a permit from the Resident Commissioner, but this was freely given. Perhaps Gibbons had not mentioned the kudu!

In his diary entry for 11 March 1896, Lugard reported the disease had reached Palapye. In the same month it appeared at Palla allegedly brought there by trek-oxen (since it followed the roads its spread does seem to have been primarily due to trek-oxen rather than game), where in May it was reported that 200 waggons were halted and 3200 oxen lay dead. Endeavours were made to stamp it out by shooting all the spans of oxen which first became infected, but the ox-waggon traffic between Mafeking and Bulawayo was so great that the few police available were totally inadequate to deal with the situation, and it rapidly passed further south.

Trekking north in May of that year on his journey to Ngami to conduct a mineral exploration of the area for the West Charterland Company, a rival to Rhodes's British South Africa Company, Lugard recorded the way was marked by the carcases of thousands of oxen putrefying in the sun, the beasts mainly dying at water holes and contaminating them. At Gaborone Lugard was told between there and Bulawayo 4000 waggons were stranded and 64,000 trek oxen lay dead on the road. Some waggons were abandoned with teams of 16 oxen lying dead in their yokes. By April the missionary Willoughby estimated £25,000 worth of goods lay abandoned on the road between Palapye and Mafeking alone. The waggons were loaded with every imaginable kind of article from ladies' stays and artificial limbs to grand pianos! Coillard reported from Zimbabwe: "Hundreds of carcases lay here and there, on the roadside, or piled up in the fields ... More than nine hundred waggons, loaded with merchandise, without teams or drivers, stood abandoned along the Bulawayo road. In a few weeks-a few months let us say-I am assured that eight hundred thousand head of cattle-some say nine hundred thousand-perished in Khama's tribe alone."

The officially reported figure was 800,000 for Bamangwato alone, Khama the chief allegedly losing a herd of 10,000; although he had sufficient left that by October he had sold off 100–200. Of his 300–400 trained trek-oxen he did not save enough to make up one span.

Altogether it was estimated 95% of cattle were lost, but it is possible the actual numbers were exaggerated. Willoughby put the Ngwato cattle population at somewhere close to a million head before the outbreak, but the national herd numbered only 1.4 million in 1970, and no doubt mass deaths from the drought were imminent in 1896 in any case.

Willoughby assessed its effect in Ngwato as follows: 60–70% of the cattle died in April; 5–10% equal to 90,000 in May–early June; 70–80% at the end of June; 70% in July, rising to 90% in September; 94–96% in October, and 96%, or 800,000, were dead by November.

The people skinned the dead animals and dealers traded in the contaminated skins spreading the disease further. They also made biltong from some of the dead animals and from those that they killed before they died, which played a significant part in the avoidance of starvation during the ensuing months. Initially 10,000 cattle were killed in Palapye by the authorities in an attempt to prevent its further spread.

Sometimes travelling at the rate of 40 km a day, about the maximum of an ox-trek, at the end of March it had reached the Notwane confluence and was travelling up the Limpopo. In August Willoughby reported Khama had cleared away all of the rubbish and dead animals from Palapye for fear of typhoid which had broken out in Bulawayo, and Ngwato regiments were employed in burning carcases to halt the spread. By August, 40,000–50,000 cattle carcases had been buried or burnt along the road between Mafeking and Shashi, but game carcases were left untouched and there is no record of how numerous these may have been.

On November 4th, in the height of a drought, it reached Ngami having spread westwards along the Boteti, and almost immediately the pools in the river beds were choked with dead oxen. Luckily in the middle of November the rains broke.

For those waggons reaching Palapye from the north, here travellers had to leave them and take on new teams. But even so, if travelling south a few km north of Mafeking there was a quarantine station where the ox-teams were shot down on the road and the waggons had to remain there. Yet in early April the plague had reached Mafeking.

Every effort was then made by the Government of the Cape to prevent its further inexorable advance southward. Several herds of cattle were shot and two lines of fences erected south of the infection, patrolled by mounted police to prevent any movement of cattle. But by October the disease had reached Kimberley. A final attempt was then made to check it at the Orange River by erecting a barbed-wire fence about 1600 km in length along the northern boundary of Cape Colony, beginning at the south-western extremity of Botswana and extending westwards to the coast.

Although this checked the speed of the advance there was an outbreak south of the fence in March 1897, and its further effects were only stopped by a successful campaign of immunization. At the end of 1898 it was under control, the last reported outbreak taking place on the Transvaal-Botswana border in August 1899.

Altogether one million cattle are alleged to have died in Botswana. One result of this was the hastening of the construction of the railway line from Vryburg to Bulawayo, which already at Palapye in August 1897 was completed in October that year. But as the trains were wood-burning, this resulted in the destruction of large trees all along the route through Botswana, transforming what was once wooded country into open bush. Another result was "fence fever". The erection of cordon fences along the borders with South Africa and Zimbabwe, later to lead to the network of cordon fences throughout the country.

7.4 Survival of the Game

Allegedly the rinderpest killed off all of the buffaloes on the Chobe river. An explorer, Percy Reid, came along the south bank from Linyanti to Kazungula in 1899 without seeing any, but there was some fairly old spoor of a small herd on the north bank, so not all had succumbed. But the German geologist, Dr. Passarge, travelling along the Boteti sometime between 1896 and 1898, allegedly found the antelope species greatly diminished.

Although large populations of susceptible animals survived in the Kalahari, the disease never re-emerged in Botswana. Unlike in other parts of Africa, particularly the Serengeti where a circular link was set up between wildebeest and cattle, as well as buffalo and cattle. This caused an annual mortality among the wildebeest known as "yearling disease" which killed 74% of the calves, and the buffalo may have experienced a similar recurrent calf mortality. The link was not broken until vaccination of the cattle in the 1960s; after which it died out in the game and led to an explosive increase of the wildebeest population.

Hodson was one of the few white men to travel through present-day Kgalagadi and Ghanzi Districts at the beginning of the twentieth century. At Xatswe Pan, halfway between Tshane and Takatswaane, he observed "In all this country there is a fair amount of large game". This was mainly gemsbok, hartebeest, and springbok.

In 1906, travelling north from Tsatsarra, or Selous's Pan, to the Linyanti, "Large game was plentiful along the route we travelled: sable, roan, kudu, wildebeeste, sassaby, pala, leechwe, waterbuck, ostrich, hippo, giraffe, eland, elephant, steinbok, duiker". All except the ostrich, hippopotamus and elephant, are species susceptible to rinderpest. The only species he did not encounter on either of his trips through the Chobe was the buffalo.

In 1907 there were "wildebeeste, sassaby, zebra and pala in extraordinary quantities" on the north-eastern edge of the Savuti marsh. The Resident Commissioner,

Ralph Williams, noted in June 1906 it was "still crowded with game to an extent which I have never seen elsewhere in Africa."

"... sassaby, blue wildebeeste, and hartebeeste are very plentiful on the Mababe and remind one of some of the scenes Gordon Cumming described. Lions are also numerous about here, and it is very seldom that a night passes in which they do not kill a sassaby or a wildebeeste."

In 1907 Hodson reported "a few eland and giraffe" around Ngerugha Pan on the old waggon road, so the highly susceptible eland had already recovered some of its numbers.

Recovery of all but the buffalo had apparently taken place in the space of ten years, despite the Tawana assertion that "a generation of children did not know game animals".

7.5 Death to Buffaloes

In spite of these other deadly diseases, it was the much less virulent viral foot-and-mouth disease which was considered most important to control and which led to the erection of the endless miles of cordon fences within Botswana. If not naturally enzootic in southern Africa it certainly has a long history there. The French traveller Le Vaillant described a disease of cattle in 1780 which "attacked the feet of oxen causing them to swell prodigiously, and often producing suppuration; sometimes the hoofs dropped off. It was not attended with any danger and generally terminated in a fortnight." Cumming also reported serious losses among his oxen from the "tongue or hoof sickness" along the Vaal River in the late 1840s, a disease well known to his local servants. Journeying to Linyanti in 1852, in the region of the Ngoha Hills Livingstone reported his oxen had swollen feet and many bled at the nose: "It is different from the Hoof disease of the Colony, for that happens when the roads are dry. Here we have plenty of water." In 1858 in a campaign along the Harts River, General Kruger reported his bullocks suffered much from the "tongue and claw sickness."

In 1892 Decle reported that between the Motloutsie river and Tuli his oxen were suffering from foot and mouth disease, diagnosed by a Dutchman who stated the transport to Bulawayo was almost stopped through it. It was in this year the disease was first officially confirmed from an outbreak in Griqualand West. The South African government veterinary officer stated it was first reported in parts of Zimbabwe that year, from where it is considered to have spread through Botswana into South Africa by means of trek-oxen. Although causing considerable calf mortality, that of adult stock was only about three per cent, and it never assumed serious proportions.

The disease forms lesions in the mouth and around the hooves, in bad cases resulting in the horny sheath separating at the heel when it can cause death due to lameness preventing the animal moving and feeding. It is a disease that although it reduces milk yields, you can live with unless you want to export cattle.

In Britain outbreaks in the twentieth century have always been controlled by slaughter, and Europe has brought in strict requirements to prevent its importation into the European Community, which became Botswana's best cattle export market because of the high subsidies paid, even if the beef was not required in Europe. The result of this has meant not only the decline of the major wild herbivore populations in Botswana through the effects of the cordon fences, but death to the buffalo as well.

Foot-and-mouth disease has been found in some 17 species of game animal in Africa, including impala, sable antelope, waterbuck, and kudu; when domestic stock have been affected (sheep, goats and pigs can be infected as well as cattle). The procedure of vaccinating domestic cattle with the virus conducted in Zimbabwe after the 1931 outbreak may have been responsible for disseminating the disease among some species of game; for while immunizing the cattle, in the game it created loci of infection.

Nine outbreaks occurred in Botswana between 1948 and 1980, with none since that year until 2002. Outbreak areas have been limited to Ngamiland-Chobe, the Boteti River and Nata; the common factor in these areas being the presence of buffaloes. Following an outbreak in the Chobe area in 1968, sampling of buffaloes, bushbucks, impalas, wildebeests and tsessebes two years later recovered virus only from buffaloes. Species which had antibodies in their blood had only minimal levels, showing they had experienced minor exposure to the disease.

In 1957 an outbreak in the north spread south into the Kalahari game herds following inoculation of the cattle with live virus. Used in Zimbabwe in 1931 this is a fairly cheap but risky method of providing immunity to acute infection. In this case the game contracted the disease from the cattle and became heavily infected, but it died out and there has been no outbreak since. No outbreaks have been recorded as originating in areas bordering the Kalahari, despite the vast reservoir of susceptible game there. Wildebeests, hartebeests and springboks sampled in 1970 showed no sign of the virus. The 1980 outbreak was in the Nata area, and it was here that it re-emerged in 2002.

There are seven known types of the virus: three occurring in Europe, one Asian and three much less severe types known as the South African Territories, or SAT 1, 2 and 3, which have never been recorded in Europe. All three SAT types have been recovered from buffalo in the north of Botswana and unlike any other animal, the buffalo can carry all three types of the virus at once, although one type predominates at any one time. In the north of Zimbabwe buffaloes have been found carrying the SAT 1 virus and yet there has never been an outbreak of this type in cattle in that region. Just as in Uganda SAT types have been isolated from buffaloes but there has never been an outbreak in cattle.

Further mystery surrounds the implication of the buffalo as a source of the disease, for although it appears to be a likely reservoir with more than half of the buffaloes in northern Botswana estimated to carry it, apparently suffering no ill effects from the disease themselves, transmission to cattle must be a rare event. It has only been transmitted experimentally to cattle by keeping them in close contact with infected buffaloes for two years; a situation which does not occur in the wild. Acutely infected buffaloes do not invariably infect other species in close proximity.

Plate 7.2 Sunset on the Chobe. Buffalo on the Chobe waterfront as they were in Selous's time

Buffalo dams transmit it to their calves, infection being acquired at three-eight months when the calves' maternal antibodies acquired at birth have disappeared, and it may then persist in the animal for up to five years. But infected buffalo mothers have failed to infect their calves. Only four weeks after infection a high proportion of buffaloes no longer excrete the virus. So normally only buffaloes in acute stages of infection, that is, probably calves, are likely to provide a potential source of infection for other animals, and then only with close contact. Although outbreaks in cattle do appear to take place when both species compete closely for water in drought periods it is unlikely to be passed via the water, but stress at such times may make both buffaloes and cattle more susceptible to infection.

As buffaloes are seasonal breeders with 90% of births taking place in the wet season of December–April, of which 65% are in mid-January–mid-March in northern Botswana, the disease may reach epizootic proportions from time to time when the calves become susceptible to infection. Mechanical transmission, for example by tick birds or oxpeckers, remains a possibility that cannot be ruled out. Where other wild species are infected, it is believed they have been infected by domestic stock which shed large amounts of the virus (Plate 7.2).

7.6 Buffalo Distribution

The distribution of the buffalo in Botswana today might lead one to assume its range was limited to the far north of the country by climate, preferring higher rainfall areas. This is not the case, although the desiccation of Botswana may play a part in limiting its present range.

As we have seen, in the nineteenth century buffaloes were found on the Nosop River, and in the 1840s on the Molopo and Notwane Rivers. Richard Orpen encountered them north of Bray inland from the Molopo River in June 1852, but they had gone from this entire region by 1872. They abounded at Kolobeng in 1847 and Oswell observed two years before this the country "of the Ba-Katla swarmed with buffaloes." Found along the Limpopo and up through eastern Botswana, Oswell recorded immense herds on the Marico River in 1845. Chapman came across a herd of 800–1000 near Serule in October 1854, and in 1865 they were encountered by Finaughty at Boatlaname. They spread westwards across the northern end of the Kalahari desert, Livingstone reporting that many drank at a waterhole south of Nxai Pan. During years of good rainfall they ranged into the northern part of the Kalahari, where as late as 1935 they came as far south as Piper Pans in the wet season, some appearing there again in the years before 1977, as well as at Khutse and Ghanzi.

In 1852 there were buffalo at Gutsha pan, they had made the water so muddy from their wallowing in July that Chapman could not drink it. In May 1853 they were to be found at Lake Xau, and in 1862 Chapman records a herd of 200 at about Matima on the Boteti and signs of spoor along the river from Kwaraga. Baldwin found buffalo on the Boteti in the region of Xhumaga in September 1859, and the following year on the south-eastern side of Sowa Pan; but of the Makalamabedi area he noted the Bushmen "drive the buffaloes about dreadfully and kill numbers of young ones every day. Yesterday a party killed 6, and the other day a Macoba killed 8 … Notwithstanding all this persecution, buffaloes are still plentiful" (Fig. 7.1).

7.7 Buffalo on the Boteti

In 1878 a group of trekking Afrikaners spent several months camped along the Boteti river and killed a great many of the buffaloes which were numerous there, allegedly driving away up the Thamalakane those they did not kill. According to Selous (1907) "After 1878 no buffalo was ever seen again on the Botletlie river", and Bryden in 1890 wrote: "The Botletlie River was a great haunt of these fine animals; but the trek-Afrikaners of 1878 seem to have hunted and driven them completely from the Lake Ngami region". But they were still there a hundred years later, suffering from the drought as perhaps they had never suffered before. I suspect some had always remained, although it is possible they could have re-colonized the area from the Thamalakane, only 90 km away, for buffaloes marked at Savuti in the 1970s moved back and forth to Hwange, a distance of almost 450 km as the crow flies. Debenham saw buffaloes and other animals at Nxai Pan in 1950, "as pictured in those early books of South African travel when the whole sub-continent was full of such game". A study of the Boteti River-Nxai Pan area in 1966 reported buffaloes at both places.

Selous found plenty of traces of them along the Chobe in 1874, and at the open valley, which might have been the Sidudu or the Nyangwe, he came across a herd of two or three hundred feeding down to the water for their evening drink just as the visitor can see them today, well over a century later. Then he considered the number

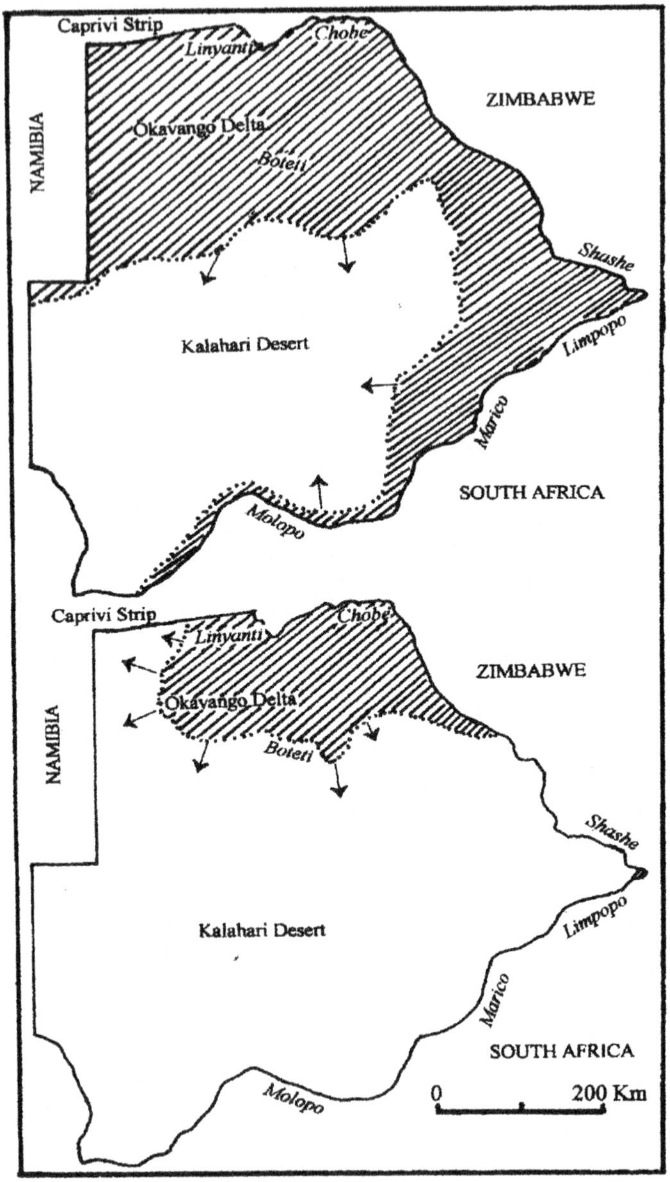

Fig. 7.1 Buffalo distribution in Botswana. Upper 1840, lower 1976

of buffaloes in this region of the Chobe was "really astonishing", and in no other part of the country that he knew of were they as numerous. Beyond Kavimba he came across an "enormous herd of these animals, lying, like the first we had seen [another large herd, closer to Kavimba], just on the slope of the sand-belt, where they had no doubt been sleeping during the intense heat of the day. This part of the country must have been utterly undisturbed by human beings for some time past, or buffaloes would never lie like this all day long, and in full view, so close to the water."

But by 1890 they had been driven farther westwards, although still to be met with in great numbers beyond the outlet of the Savuti river. Once very common throughout the country around Lake Ngami, by 1892 it had entirely disappeared from there.

In 1876 Selous shot four buffaloes on the upper Ramokgwebana, "which are about the last that have been shot in that part of the country."

7.8 Uncompromising Destruction

But after the onslaughts of the Afrikaners, possessed of the same mad lust for slaughter as the hunters of the American bison on the Great Plains, came the demands of the livestock industry, requiring the uncompromising destruction of buffalo wherever livestock occurred, and their separation from stock.

Due to the implication of the buffalo in harbouring a reservoir of foot and mouth disease, in 1982 a "buffalo fence" was constructed around its habitat in the Okavango swamps to comply with the European Community requirement. Two cordon fences are required, one to limit the carriers (the buffalo), and another to separate the (vaccinated) cattle population from those which are not vaccinated. The aim was to build up a network of fenced cells throughout Botswana so that disease outbreaks, not only foot-and-mouth but other possible contagious diseases also, could be isolated and treated.

In 1961 the Director of Veterinary Services announced: "it is chiefly because of the dangers which game afford in the transmission of this disease [foot and mouth] that we are interested in bringing about a rapid reduction in the game population, particularly in those areas of the country bordering upon the swamps of Ngamiland, Zambesi, Chobe and Botletle Rivers where the disease appears to be maintained in an enzootic form in game animals and from which it is transmitted to cattle at more or less regular and monotonous intervals."

It is a policy the Veterinary Department would still like to see implemented.

7.9 The Tsetse Fly

Buffalo became implicated in another disease, one fatal to both man and cattle, that of sleeping sickness in man and *nagana* in cattle, the protozoan blood parasite transmitted by tsetse flies. It was well known to locals in southern Africa who avoided the infected patches of woodland where the tsetse fly occurred: away from river valley thickets and the Okavango Delta.

Although there were tsetse flies in the Tswapong Hills in Cumming's time, in 1867 tsetse fly country was considered to begin north of Pandamatenga at the Matetse river, so traders and hunters visiting the Victoria Falls left their waggons and horses at Pandamatenga and proceeded on foot. Westbeech, who kept a store at Pandamatenga on a hillock overlooking the most southerly of the headstreams of the Matetse River, by May 1886 was able to consider building a new store at Kazungula at the junction of the Zambezi and Chobe rivers due to the retreat of the fly. Coillard reported in 1884 that it had retired "following the buffalo in its flight before the hunter's rifle", and was able to camp 16 km from the Zambezi. He noted the fly was ever tending to recede from the parts it formerly haunted, following, he claimed, the buffalo in its migrations.

In 1850 Livingstone was informed the tsetse fly abounded on the banks of the Thamalakane. When he reached the Chobe he found its habitat was sharply delineated, for although the south bank was infested, the northern bank only 45 m distant, apparently did not harbour a single specimen and cattle could be kept there. "This was more remarkable, as we often saw natives carrying over raw meat to the opposite bank with many tsetse settled upon it." Livingstone reflected "it is probable that with the increase of guns the game will perish, as has happened in the south, and the tsetse, deprived of food, may become extinct simultaneously with the larger animals."

When he first made his way to Linyanti in 1853, after leaving the Goha Hills he passed up the Kataba valley to the Chobe river, which apparently was the only gap "in the whole country destitute of tsetse."

In Zambia he was led to suspect the tsetse had some connection with the presence of elephants, and noted that the Portuguese in the district of Tete in the east called the fly the *musca da elephant* or "elephant fly".

Selous wrote that although he had hunted elsewhere in tsetse-infested countries in 1872–73, never had he met the fly in such numbers as he did along the Chobe River in 1874–76 during the months of September and October. "I think that this plague of "tsetse" flies, along the Chobe and Zambesi, is due to the enormous numbers of buffaloes that frequent their banks, as they always seem very partial to these animals."

It was Selous who stressed a supposed connection between the presence of tsetse fly and buffalo, and pointed out that in a very few years after the complete extinction of buffaloes in certain areas, the tsetse flies ceased to exist. For years after the disappearance of both buffaloes and tsetse flies from the valley of the central Limpopo, other game such as zebras and antelopes continued to exist in considerable numbers, so Selous drew the conclusion the flies could not maintain themselves on the blood of other species.

Up to 1878 there were two fly-infested tracts of forest to the west of the Boteti, but soon after the buffaloes allegedly ceased to exist here, so did the tsetse flies. In the early 1870s the flies swarmed along the southern bank of the Chobe, but farther east nearer to the Victoria Falls both flies and buffaloes disappeared, the buffaloes driven out by hunting. Selous could find no tracks of buffaloes at the Zambezi in 1877, but had found them there three years previously. On the Chobe Selous recorded that the flies continued to haunt the southern bank in great numbers long after the buffaloes had ceased to live there constantly, although they still visited in the rains. By 1885 only a few flies existed between Leshomo and Kazungula and there were no buffaloes there. A few flies still lingered in 1888 and later in that year they increased rapidly in numbers in August and November, but the next year just as rapidly diminished in numbers and then died out. In 1899 no flies were to be seen between Linyanti and Kazungula. They are believed to have remained absent from the region of Kasane until 1940, and increased noticeably after 1952–3. No buffaloes were seen there between 1940 and 1946, but numbers then increased considerably, probably because of the disappearance of cattle from the area.

In the early 1850s tsetse flies existed in great numbers along the Thaoge river, allegedly later dying out. But Lugard in 1899 mentioned a residual focus of many tsetse to the west of the Savuti "near the main channel of the Chobe River", despite the absence of buffaloes. In 1928 game was rare in the vicinity of Maun but the tsetse belt was expanding rapidly, getting closer to Maun each year. It was then within six km and had spread over some of the best grazing land that the Tawana possessed.

Although in recent years it has been established that tsetse flies are attracted to a constituent of buffalo urine, the retraction and expansion of the tsetse fly's range resulted from varying climatic conditions, the fly dying out when its habitat becomes too dry. It had in fact retreated from much of its southern African range several decades before the rinderpest panzootic, but this circumstance was studiedly ignored.

7.10 Eliminating Game

Unfortunately the spurious association between the disappearance of the fly in the drought years towards the end of the nineteenth century and the destruction of game populations by the rinderpest, led to the promotion of game elimination programmes to control the tsetse. In Botswana the first attempt to control the fly by this method was a game drive held in 1916. Then in 1939 the Maun region was thrown open to hunting and the following year local tribesmen recruited as hunters. Between 1940 and 1941 a total of 1887 head of game was shot, mainly along the Nhabe river. Hunting was properly organized in 1942 following Dr. Macaulay's recommendations, and continued until 1967, a system of parallel fences being constructed from 1945 within which all game was to be exterminated.

Between 1942 and 1964 a total of 60,638 head of game was shot and an unknown number wounded. Of this total 8635 head were buffaloes and 8479 kudus, but the most numerous were warthogs, with 10,395 destroyed. Wildebeests totalled 2827. Although the numbers of giraffes killed were concealed under "others", they were specifically sought to keep them away from the fences on account of the damage they did to them. Only three European hunters were allowed to shoot them and policy was to shoot just one in a herd to attempt to frighten the others away.

Analysis of the kills suggested the killing was not reducing the number of buffaloes coming into the area and that there may even have been an increase. Only tsessebes, wildebeests, and kudus showed a decline in numbers shot and this is thought to have reflected a general decline in these species throughout Ngamiland. All other species showed an upward trend in numbers being shot so the method was not effective in controlling the tsetse fly although reducing the numbers of game animals.

Around 1990 an estimated 20,000 buffaloes were resident in the Okavango Delta, as well as seasonal migrants from the north. The total for northern Botswana was put at over 32,000. In 2014 the total was estimated at almost 27,000, down from 60,000 in 2006. But we are fortunate indeed that Dr. Macaulay's recommendations for the total destruction of game were not taken up. Let us hope that Botswana will keep it that way, striking a balance between cattle and wild life, not confrontation as has been the case in the past.

Chapter 8
The Expiring Kalahari

Abstract Drought cycles recurrent phenomenon since Livingstone's day. Historic droughts. Drying of waterholes, rivers and streams in nineteenth century. Lake Ngami fluctuations. Hippo and silting of Okavango Delta. Makgadikgadi salt pans an ancient vast lake. The Savuti another fluctuating watercourse. Mababe flats alternate between marsh and dry plain. Resumption of flow not due to rainfall. Fall of South African water table. Temperature rise in Botswana. Removal of vegetation cover, overstocking and overuse causes of increasing dryness.

Keywords Botswana · History · Drought · Ngami · Savuti · Mamabe

8.1 Unprecedented Adversity

Gaunt reedbuck with frightened eyes crouched in immeasurable misery, desperately trying to cling to life in the scant shade of the almost leafless bushes fringing the plain. The tsessebes suffered equally, standing hunched, listless and pathetic, bones protruding through their dusty hides, their brilliantine-like sleekness gone.

The wide grassy plains of the Mababe Flats host an exuberant profusion of animals when the Savuti Marsh holds water, now an ephemeral phenomenon after the rains. Lions are in abundance, although today the prey is largely seasonal. Impalas remain, but the herds of zebras and wildebeests migrate north to the Linyanti as soon as the dry season sets in. When last I saw the area it was in the throes of drought and I suspect the reedbuck population succumbed to the point of extinction. Already a remnant of their former numbers, neither could many of the tsessebes have survived.

Prone to a 20-year drought cycle, from 1981 to 1987 Botswana experienced one of its worst droughts on record. although worse was to come. Past references show that severe droughts have been a recurrent phenomenon since Livingstone first visited the country in the middle of the nineteenth century, and the droughts witnessed since that time, both in Botswana and throughout southern Africa, have led to the belief that the country was drying up. Most significantly, the vast inland lake of Ngami, which

C. Spinage, *Big Game of Botswana*, https://doi.org/10.1007/978-3-031-60033-3_8

Plate 8.1 Reedbuck suffering from drought

Oswell and Livingstone were the first white men to set eyes upon, had disappeared within 32 years of their discovery.

Livingstone observed that droughts in Botswana were reported to occur in cycles, "How many years are required to complete a cycle it is impossible to know", he wrote, but he believed it to be 11 or 12 years, a cycle experienced at the Cape. Prior to Livingstone's arrival, 1839 had been a remarkably wet year which re-charged many dried up fountains (Plates 8.1 and 8.2).

Drought was recorded in South Africa in 1777–78 and in 1803, and the years from 1814 to 1824 were a time of devastating droughts throughout. Further droughts followed in 1834 and 1844–47, that of 1847 being particularly disastrous. The winter of 1851 was preceded by severe drought in Cape Colony in which farmers lost two-thirds of their stock. Drought in 1862 was widespread; and again in the early 1880s. The Ngwato harvest failed in 1876–77 with famine the following winter. To buy food, Khama gave his people the proceeds of the yearly hunt of feathers and ivory worth from £2000 to £3000; at least £60,000 to £90,000 in today's value.

In 1919 the Zambezi fell so low that one could almost walk across the Victoria Falls and the worst drought on record occurred; although that of 1861 was said to have been as bad. The years 1923–1929 were drought years, but then again at Maun 1932–3 was recorded as the "lowest rainfall in history". November 1935 was reported as the worst drought for 20 years in the north and north-east, and cattle were dying at the rate of 1000 a day. By 1938 it had spread to the south and by that year the cattle herds were reduced by about one third from an estimated 1,000,000 in the country. In 1945 conditions were again so dry it was possible to drive across the valley at Kachikau without wetting the tyres of the car.

Plate 8.2 Tsessebe suffering from drought

But some considered there was no reliable evidence for any great climatic change in South Africa since Europeans first settled there in 1652. Unusually heavy rains in 1879 caused the Molopo River to flow, as they did again in 1896. The years 1906–7, 1914–15 and 1917–18 were also years of heavy rainfall, while intervening years were low. A good year was experienced in 1917 and exceptional floods caused the Molopo to flow once more. It flowed again in 1933–4, a year when the Nosop, which had first stopped flowing about 1806, also briefly flowed. The whole of the Linyanti to Chobe area was like a vast inland sea and all the swamps were flooded. The Molopo flowed again, reaching Werda in 1967, 1974 and 1988. The Nosop also flowed in 1971 and 1974. Yet there is no doubt some rivers in Botswana have ceased to flow since the beginning of the nineteenth century, and many springs and waterholes have dried up. This has put pressure on both man and wild animals and resulted in a drastic and continuing diminution in range and numbers of game. Unremitting slaughter, disease, and drought, have all combined to produce unprecedented adversity for the wildlife.

8.2 Extraordinary Drought

The rainy season in Botswana is normally from October to April, but rain can only be depended upon from December to March, with sometimes a little in April. The remaining months are dry. Thus when Dr. Andrew Smith reached the Notwane in

September 1835 and found it to be a series of small pools, this is what we would expect to find today, "more mud than water" as it was described in 1896; and in September of that year "no better than a feeble brook, flowing slowly."

In the dry season of 1826, a few km east of Bray Bain recorded that the Molopo was dry except for a few muddy pools, but after the rains near Ramatlabama it was a stream "thirty yards wide."

As with so much else, we owe our most detailed descriptions of early droughts to Livingstone. He had installed himself at Kolobeng in 1847 and recorded that in the second year "again no rain fell. In the third the same extraordinary drought followed." Not ten inches of rain fell during the two years and the Kolobeng ran dry, "so many fish were killed that the hyaenas from the whole country round collected to the feast, and were unable to finish the putrid masses." A large crocodile was found high and dry among the victims. Very little rain again fell in the fourth year. Very great numbers of game; buffaloes, zebras, giraffes, tsessebes, hartebeests, wildebeests, impalas, rhinoceroses, and others; concentrated at some springs near Kolobeng and were exploited by means of pitfalls. On his arrival he noted it seemed to have been the haunt of everything wild. Hyaenas abounded, buffaloes were in immense herds, and zebras were quite tame in the thickly wooded country around. There were tracks of elephant also.

In January 1849 he wrote from Kolobeng, "This is the fourth year of stinted crops, and the Bakwains formerly were remarkable for the abundance of their corn … All the fountains in the country are dried up, so we cannot grumble and wish we had been elsewhere in this region. I have never seen anything like this drought … we hear that there are some tracts of the Bakalahari country in which the trees are killed". And, "The drought in the Kalahari side has been so great very few have been able to go thither in search of skins, so they are scarcer this year than we ever knew".

"All the streams and fountains in the tract of country inhabited by the Bakwains and Bangwaketse have been subjected to the same cause of desiccation. The chief of the Bangwaketse lately showed me [1851] nine streams which were perennial when Mr. Moffat visited Makabba, but now during the greater part of the year they present the spectacle of dry watercourses."

While it appeared that Kolobeng at first missed the rain which fell in adjacent areas, the last of a series of dry years in 1850–51 exceeded all the preceding droughts in the extent of country over which it prevailed, from Shoshong to the Lake Ngami region, as well as Bakwena.

Lephephe, where in 1844 Cumming encountered zebras, wildebeests, tsessebes, impalas and springboks; in June 1849 "afforded another proof of the desiccation of the country." The first time that Livingstone had passed it in 1842 it had been a large pool with a stream flowing out of it to the south, "now it was with difficulty we could get our cattle watered, by digging down in the bottom of a well." In mid-June 1854 Robert Moffat described Lephephe as "a valley, or rather what is more properly called a *pan*, for there is no outlet. There is a deep hollow in it, covered with reeds, on the outer skirts of which there are pits, but at present all filled with water, so that the oxen can drink without trouble." In mid-November the water was "not very plentiful nor good."

In October 1851 the country round Kanyu on the Maun road contained a great number of springs in limestone rock. The dry "Bushman's Pits" is a reminder of this today, then: "A considerable number of bushmen live in the vicinity and they seem to have abundance of food. Leaving this district of springs … we crossed an excessively dry and difficult tract of country". In April of that year of the area farther east around Odiakwe, Livingstone wrote game abounded and the people were reported to follow it in its migrations and live on the zebras and wildebeests and other game as if they were their cattle: "all the Bushmen were strong well fed looking men." There were so many of the limestone springs to the east that the area was known as Matlomaganyana or "Links", as of a chain. The water in the springs was independent of the rain.

But "The drought had been greater this year [1851] than during any of the years preceding it since we came into the country."

Today Bushman Pits is simply a small dry hole in the limestone with rubbish in it. In 1946 the supply of water was described as "uncertain", but in 1949 the wells still yielded a good supply of water at a depth of seven meter. In 1950 a cattle post was there and although the area was open grassland such as is now found farther east, for nearly two km around the grass was eaten or trodden out replaced by inedible weeds. In that year the Colonial Development Corporation took over the cattle post, which at times was to hold 10,000 head of cattle. Within 20 years overgrazing had changed the area to the dense thorn thicket characterizing it today.

Travelling eastwards from Gobabis in Namibia almost to the Botswana border, reaching as far as Rietfontein, in September 1851 Galton wrote of the drought: "The dryness of the country was now really alarming; all the watering places that remained were crowded with cattle, and every blade of grass within miles of them was being eaten off. Over a great part of Damara-land rain had not fallen more than ten times during the whole rainy season; and a mortality from starvation had already begun among the cattle".

In April 1851 Livingstone could get scarcely any water from the wells at Boatlaname, and there was none at Lephephe. By January 1852 there was none at Boatlaname. Yet in September 1851, while he was at Letlhakane, there were three days of rain which made the Mokoko (= Letlhakane), a dry river bed, assume the appearance of a large river: "The old people remember the time when it flowed so through the whole year". In February 1853 when he passed down it again it was dry once more. Later he passed "the small well named Orapa", little suspecting the wealth of diamonds hidden in the area.

Bryden recorded in 1890 the Letlhakane was now always dry, as were the Letlhakane pits, a former water-hole.

But in 1852 there had been a larger fall of rain than usual which allegedly was the completion of a cycle of 11 or 12 years, the same phenomenon being reported to have happened on three previous occasions back to about 1820. At Chukutsa in February 1853, Livingstone reported the rainy season had been delayed beyond the normal time, but in the area of Nxai Pan "Instead of the dreary prospect around Koobe and Nchokotsa, we had here a delightful scene, all the ponds full of water, and the birds twittering joyfully." But to the north-west, although the rains had been "copious", now great numbers of pools were already drying up.

At the end of the dry season in November 1854, Robert Moffat found very little water at Boatlaname pits and travelled on to Kopong, his oxen having gone three days without drinking. Here also there was very little and they had to dig for it. But thunderstorms were beginning and unusually heavy rains fell during the 1854–55 wet season, although according to Green, in 1855 neither the Boteti nor the Zonga (= Thamalakane?) rivers had been flowing for some time. Travelling in the wet season south of the Makgadikgadi Pans, in the region of Orapa along the line of the present road, Chapman reported two feet of water in January 1854. From Tsokotsa he was not able to get out of his waggon until he reached Kgokonyane, but Lake Xau was apparently "a reedy swamp".

There were good rains again in 1855–56, but in 1857 the river at Shoshong was dry and the Shashe river was dry in October. Then the year 1859 saw an "unprecedented" drought year, "even as far as the banks of the Zambezi" according to Moffat. Baldwin reported that no rain fell for ten months at Shoshong, but then there were four days of rain at the end of October. The year following was a good year according to Baldwin, and he found plenty of water in May except at Masowe. But Chapman wrote in January 1860 that in the whole country up to Lake Ngami the springs had failed and "if the desiccation continues a few years longer at the rate it has been doing the last four years, I fear we shall only be able to reach Ngami during the rainy season … we had to dig at Koobye, Ghanzi, Gunegga and other large springs, where a few years ago hundreds of elephants, rhinoceros, giraffes and large herds of smaller game drank during the whole dry season. They have now dried up so much that scarcely a kettle of water can be got for coffee."

In September 1861 all the waterholes at Kobe (near Kuke) were dry. It had not rained there the past year and water was very scarce. Chapman's companion Thomas Baines wrote there was "no game of any kind about the place, though Chapman has never known it fail thus before." And: "Formerly this place was a vlei or small lake, at least one hundred yards across, and even on Chapman's last journey there was an ample supply for his oxen as well as Lamert's."

"The Botletlie too, they had told us for years, became in its lower course a broad and deep river abounding in hippopotami and crocodiles. The fallacy of these exaggerated reports at once became evident on my visit to the spot … the Botletlie gradually widened into an extensive salt basin [the Makgadikgadi salt pans] … which is generally dry for half the year or more … the older Bushmen assert that about thirty or forty years ago it never dried up (1810–1820), and abounded with hippopotami, crocodiles and fish. But suddenly, they say, the waters from Lake Ngami ceased to flow; the lake dried up, and the dead fishes and animals were devoured by the vultures".

In 1862 Chapman wrote: "When I first entered this country I found many of those ponds with an abrupt bank all round, and the water then, as was usual, nearly up to the top; but even in ten years a wonderful change has taken place; the water has gradually diminished, owing no doubt to the general desiccation going on; and in places where formerly I could swim we now have to go underground for a supply of water. Whenever late and heavy rains fall the natives say that things are very much better, though never as it used to be of old. They say the country is dead."

In 1862 the drought was widespread, followed at the beginning of 1863 by good rains in the interior, "but we long for more still" wrote Elizabeth Price at Kuruman in February. Chapman remarked in that year: "About 20 to 30 years ago there was a large swampy stream at Ghanze, where one might now die of thirst if the Bushmen did not open the pits every year".

In August of 1871, Macheng, chief of the Bamangwato, was to write to Mackenzie from Shoshong: "there was but little rain and the corn is scarce in consequence. The veldt is very bad and some oxen are dying … Game is very scarce; those that have gone hunting have suffered much from hunger."

8.3 Where Is the Water Now?

Gillmore made frequent references in 1876 to the drying-up of Botswana. Writing of Shoshong he noted a dry river bed ran through the centre of the site: "In the memory of some of the oldest inhabitants it was a constant stream; but as in many other parts of South Africa, more especially in the country bordering the Kalahari, it has gradually dried up. The hills which form the background of the town are very rugged, barren, covered with loose stone, and almost destitute of vegetation." Leaving Shoshong south of the hills and travelling east, he came to the first watering-place: "a large pool in the dry bed of a river [Bonwapitse]. Not many years back water also constantly flowed here, but, as at Shoshong, it is rapidly disappearing." Masowe, Cumming's "Elephants' own Fountain", was still "never without water", and still a favourite haunt of lions, never lacking a family frequenting it; although Baldwin had found it dry in 1859. Of the route from Masowe via Lephephe to Molelopole, Gillmore wrote: "It is strange that evidences of numerous springs and vleys that existed in former days are constantly to be found; and old Bechuanas have assured me that they can remember that they always contained water, But where is it now?"

When he had crossed the Okwa valley in 1852, McCabe had remarked "This valley … in many seasons holds water", an unlikely event today.

Anderson recorded drought conditions in 1877, when the Afrikaners trekked westward and suffered great hardship. In the north, and probably therefore elsewhere, the summer rains failed in 1878. Selous, travelling north-west from Nxai Pan (which was dry) to Mababe in May 1879, having found water at Bushman Pits which was thought to be a permanent spring, then found almost all the pans had dried up between there and the Mababe river. The last remains were found at one pan two days' journey further, and then he did not find water again until he reached the edge of the Mababe flats, where there were pans with water in the fringing mopane woodland, probably the Chosoroga and others.

In September 1879, north of the present border in the Linyanti region, Selous recorded reaching a pool of water in the bed of "what was evidently once a river". It had been a river at the time of Livingstone's visit to the area, one of the network in the swamps north of Linyanti it flowed from the Chobe to the Zambezi. In the same year Gillmore reported that a river which had flowed near Molepolole four years before, was now almost dry.

The rains apparently failed in 1881–82, for a long-protracted drought is reported at Shoshong in March 1882. But Schulz and Hammar reported that in 1883 the rainy season was exceptionally heavy, at least in southern Angola, for in July 1884 the Chobe river was over four meter higher than usual. Selous also reported exceptionally heavy rains in 1883–84 with an extraordinarily heavy rainstorm in May 1884, "a full six weeks after the end of the usual close of the wet season."

These rains may have served to fill the Ncamasere in the extreme north-west, in 1884 a stream flowing into the Okavango but was now dry.

Farini referred to a long drought in southern Botswana which had "driven the game away" but recent good rains in 1885 had resulted in plenty of tsama melons and the game was coming back in abundance. Leaving Khuis on the Molopo river "Game now became plentiful, especially gems-bok, which we saw oftener than anything else".

By 1886 Bethell was to write: "Rivers in Bechuanaland mean the dry beds of rivers, which fill during the rains for two or three days, and then dry up again."

Traditionally during droughts a Tswana family might move to the hunting-field, living on game and wild roots. Mackenzie pointed out in 1887 that as the game had disappeared, the native man took himself to the Colony to seek work and food. They would say: "It is of no use for us to go out with our guns any more; the game has disappeared; to go on the hunt now is to go and work for the white men." Thus in years of extreme scarcity, when they would have shut up their house in the town and removed to the hunting-field, the whole family now went to neighbouring European villages or farms to earn wages for food. Droughts and the destruction of game thus led to the beginnings of the migrant labour system that was to play such a significant role in the Bechuanaland economy.

In 1890 Decle records the rainy season had been "one of the worst for many years, and scarcely any rain had fallen". Bryden also reported that in the early part of 1890 there were short summer rains and a prolonged drought. The Molopo was nothing but a desiccated channel: "Nowadays no water flows in the Molopo, even in time of rain, for more than thirty miles beyond Mafeking. So also with the Nosop". It last dried up completely in 1894. The Nosop was dry in 1906 and had a number of trees growing in its bed when Hodson visited it.

In 1895–6 there was again a severe drought.

The next century rains were more than usually heavy in the Angola highlands in 1924 and the water in the swamps had not drained away before the 1925 floods began, although rains were low that year in Angola. The Chobe then reached back to the Okavango along the Selinda spillway and Kabulabula was all but submerged in a great flood on the 9th May 1925.

8.4 The Story of Lake Ngami

Lake Ngami gets its name from the Yei word for giraffe *n'ghabe*, thought in its turn to come from the word *n'ghabba*, meaning to rock. A reference to the giraffe's rocking motion when galloping. One theory is that the lake was so called because of the large numbers of giraffes which occurred there, another that the waves on the lake "rocked". The Tswana name for the lake and the Boteti flowing into it, is said to mean "the water of the giraffe."

It was on the 31st July 1849 that Oswell, Livingstone, Murray, and Wilson, reached the fabled lake: "on the 1st of August, 1849, we went down together to the broad part, and for the first time, this fine-looking sheet of water was beheld by Europeans ... It is shallow, for I subsequently saw a native punting his canoe over seven or eight miles of the north-east end; it can never, therefore, be of much value as a commercial highway. In fact, during the months preceding its annual supply of water from the north, the lake is so shallow that it is with difficulty cattle can approach the water through the boggy, reedy banks. These are low on all sides, but on the west is a space devoid of trees, showing that the waters have retired thence at no very ancient date. This is another of the proofs of desiccation met with so abundantly throughout the whole country."

It was next visited by Frederick Green in August 1851. Then Andersson reached it from the west in July 1853: "on making a nearer survey of the Lake, I experienced some disappointment as to its attractions." Its average breadth he estimated at 11 km, with a maximum of 15; and its circumference about 100 or 113 km: "there are unmistakeable proofs of its having been at one time of smaller dimension than at present; for submerged stumps of trees are constantly met with." He attributed this to a sudden and unusually large flood. The locals spoke of its waters retiring daily "to feed", and he confirmed there was a considerable daily tide, receding during the night and returning to its maximum level at about nine o'clock in the morning.

Chapman reached it in November 1853 and recorded it as receding: "As the lake is receding, the fish become plentiful, and waterfowl of every description flock hither".

It was probably not as shallow in the centre as Livingstone thought for the canoeists would have kept to the shallower edges avoiding deep water, although it would have fluctuated in depth annually as it was fed from the Okavango Delta mainly via the Thaoge river: "Round the lake is a belt of plain covered with a short prickly grass, which generally grows on soil emitting a saline efflorescence. This plain was formerly covered with water, but during the last twenty years the waters of the lake have receded very much, generally for two or three miles all round". In November 1855 he was sorry to find it was becoming alarmingly shallow, with an average depth of three and a half feet of water. The locals told him that it had "died" not long before Livingstone's arrival there, and in 1863 he placed the date at about 20 years before, as men of 27–28 could remember paddling canoes in areas now dry. While there he saw the dry reeds ignited by a flash of lightning.

In 1863 he reported: "The Teouge [Thaoge] River is said to be dying, which is the reason that the Lake cannot fill. Formerly it entered the Lake by many channels, which are now dry watercourses, and the intervals between, formerly always under water, are now grazing grounds of hundreds of wild animals. It seems to have spread over the plains to a great distance on either side and formed one continuous sheet of water like the Lake today …".

"Within the lifetime of some of the Lake people and Makobas the Ngami has gradually receded a mile or more all round, and within the knowledge of white men still living fountains everywhere have been drying up. I have had abundant opportunity of noticing the same thing going on gradually during the last ten years. The natives coming from Lebebe also insist that one branch of the Teouge (or Okavango) diverges towards the west coast".

But as is usual when a wet or dry phase is ending, the climate tended to be erratic, and at the beginning of 1862 the harvest at Lake Ngami was very good. In the inhabitants' terms they had killed a *namani-tona-tlou*, a great bull elephant calf.

According to tradition, before about 1760 there was no Lake Ngami but a river running through the plain covered with leadwood trees *Combretum imberbe*. Then in 1760 the Thaoge River flooded the plain and by 1813 a "great water" was reported. About 1880, 17 years after Chapman's misgivings, the Thaoge supply dried up, but may have continued to flow sporadically for a few more years. In August 1877 McKiernan reported the lake to be "a fine sheet of water" abounding in hippopotamuses and crocodiles, although he also wrote of poling almost all day on the shallow eastern side and seeing only one hippopotamus. The Nhabe River had a good strong current into the lake but by 1881 this "great water" was dry from "end to end" according to the missionary Hepburn.

In 1886 it was full again but very shallow and salty. Nicolls reported it as having been exceptionally low in 1889, but in 1890 there were a good many hippopotamuses in it. Four years later it was dry once more, the Thaoge, which had flowed again, ceasing in 1895. In 1895–96 there was again a severe drought but the rains broke in the middle of November 1896, a year when there was not a drop of water in the former lake and its bed was covered with reeds. However, although the Thaoge has never since flowed into Ngami, the lake's other feeders, the Kunyere and Nhabe rivers, continued to flow and the lake bed alternated from wet to dry over the years. It was full in July 1899 following unusually heavy rain, but six months later dry again.

In 1906 Hodson recorded "The once famous Lake Ngami now presents an appearance of a vast and reedy marsh intersected by streams, no large expanse of water being visible from its shores … The surface of the lake is covered with a crust which gives way at every step, precipitating the unwary into unknown depths of mud or water."

According to the Commissioner Ralph Williams, whom Hodson was escorting, the locals said that there was a great deal of water in the middle of it at that time (1906), although Hodson thought it to be quite dry and could see nothing but an expanse of reeds. He recorded seeing many varieties of game on the western side, "chiefly blue wildebeest".

It was known in 1887 that water coming down the Okavango was lost in the extensive swamps of the Okavango Delta, although some flowed down the Thaoge into Lake Ngami but was insufficient to keep it full. When there was flooding excess water flowed to the Chobe via the Ngogha in the north of the Okavango, which normally flows into the Thamalakane and on into the Boteti, but does not reach Ngami. At one time a branch of the Ngogha, the Mababe, flowed to the Mababe flats, but this ceased in 1910.

The cessation of the Thaoge was attributed by Passarge about 1899 in part to blockage by papyrus caused by the abandonment of papyrus rafts, used by the Bayei from the beginning of the eighteenth century. These rafts are simply bundles of papyrus tied together which are abandoned after use. Blocking the channels the papyrus then puts out roots and grows into a mat of vegetation. In August 1889 Nicolls found the bed of the Thaoge choked by rotten vegetation and papyrus roots, with "the river slowly percolating through the mass". But the principal cause is silting-up, which then allows the papyrus to get a hold, the sedimentation rate giving a channel a life of about 100 years.

In 1909 there were exceptionally heavy rains, but it was in 1914–15 and 1917–18 that it was fuller than it had been "for a long time", as these were years of heavy rainfall over South Africa. In 1922 it was hard turf, caused by the natives burning off the reeds in the dry seasons, the change to grass taking place in less than four years.

Following good rains in the Angola highlands in 1924, the following year a river ran through the plain again, a record flood year in the Okavango Delta and the Boteti River flowed beyond Mopipi, which it had last reached in 1919. If the 1925 rains had been as good as those of 1924 the river would have run once more into the Ntwetwe Pan. In 1925 there were still immense tree stumps along the line of the river bed of the Mopipi channel showing that it had once flowed frequently.

In 1928 Lake Ngami was somewhat less than one third full, and on two occasions in recent years before this exceptionally heavy floods had come down inundating up to one-third to one half of the lake area. In 1951 it fluctuated from dry to 30 km in extent, and back to dry again in one year.

The lake bed has been dry for two or more consecutive years eight times in the past century. Maximum levels higher than Livingstone witnessed appear to have been attained in 1898–9, 1904, 1925–6, 1968–9 and 1978–9; at least in 1925–26. Between 1974 and 2016 there were three floods and then after reaching another high level it dried out in 2019 until in late 2021 it filled again covering some 278 km^2. But whereas the fluctuations in level appear to coincide with the 20-year drought cycles, there is no direct correlation with rainfall in Ngamiland, nor with the inflows from the Okavango which derives its water from the Angola highlands. Because the lake level has fluctuated just as much last century as it apparently did during the last two, and because it has apparently reached higher levels the century before last, it is considered these fluctuations in themselves offer no evidence for increasing desiccation of the country, contrary to Livingstone's supposition. Yet desiccation there has been, and by 1925 all the great baobabs *Adansonia digitata* "but one" that Oswell had seen on the banks of the Boteti River in 1850, had gone. As had all the hyphene *Hyphene petersiana* palms except at Lake Xau.

Geological evidence suggests the region has been subject to phases of desiccation in the past 2000 years, with the final death of Lake Ngami occurring in the last three centuries.

8.5 A Hippopotamus Hypothesis

Schwarz suggested in 1925 the silting-up of the Okavango Delta channels was due to the hippopotamuses having been largely exterminated. They walked up and down the channel bottoms and kept the sand stirred so that it shifted in the feeble current. They also played an important role in forcing passages through the papyrus and are alleged to have opened up the Ngogha in this manner about 1850. Before this time it was said to be only a shallow swamp channel, then hippopotamuses broke through in great numbers and trampled a hippo path eastwards. Large pools were then formed by the hippopotamuses hollowing out breeding-places.

Hippopotamuses were numerous in the Ngogha in 1921 and very aggressive. In Livingstone's time the Chobe River was "much infested" by them, and although there is no similar observation at the time for the swamps, which were largely impenetrable, perhaps the destruction of hippopotamuses after the introduction of firearms may have led to accelerated changes taking place there. Selous considered hippopotamus to be still abundant in the Chobe in 1874.

8.6 An Ancient Lake

Livingstone realized immediately in 1849 the Makgadikgadi salt pans were the remains of an ancient inland lake, and Lake Ngami he considered to be the last puddle of this once mighty lake before the Zambezi, which had once fed it, deflected eastwards. E. H. L. Schwarz, a controversial South African professor of geology who devised a plan to open up the channels in the swamp believing that it could make the Kalahari green again, deduced there had been an ancient "Greater Ngami", which once covered the Okavango delta and Ngami, 77,700 km^2 in extent, almost 12% greater than Lake Victoria. The Makgadikgadi was 38,850, or 18% greater than Lake Tanganyika. If these once great lakes were filled with water again, he argued, the whole climate of the area would be altered, making the Kalahari once more a wetter place.

More recent studies confirm the former existence of this giant lake but not his predictions of a greening of the Kalahari. Lake Palaeo-Makgadikgadi embraced the Okavango Delta, Lake Ngami, the Mababe Flats and the Makgadikgadi Pans; covering a total area of 37,000 km^2. Its ultimate drying up following its highest level about 40,000–35,000 years ago, probably followed tectonic movements along fault lines of the earth's crust in the Okavango Delta. These movements still take place today. Twenty-two earth tremors were recorded in the Delta region in one year between 1952 and 1953, with a significant earthquake in Maun in 1952. In a nine-year period ending 1974 there were 38 tremors, and this level of disturbance still continues. At any time we might expect other significant changes to water flow regimes in the area influenced by these tectonic events.

8.7 The Savuti

Another watercourse to show intermittent flows is the Savuti. First described by Livingstone in 1851: "The Souta was about 3 feet deep and about 40 yards broad", it flowed eastwards from the Linyanti into the Savuti Marsh at the northern end of the Mababe Flats, but had stopped flowing to there by 1879 and the channel dried out sometime in the 1880s, perhaps 1888. The Savuti Marsh was described in 1910 as "a plain surrounded by forest and forming a reedy marsh in its centre, this marsh at its southern extremity gradually resolving itself into three or four well-defined streams, which, flowing south, join one another as they leave the marsh, and together form the Mababe River." In 1925, from the age of the fringing mopane *Colophospermum mopane* and thorn scrub, Schwarz considered the Savuti Marsh had been dry for a very long while, longer than Lake Ngami or the Makgadikgadi.

In 1958 the Savuti mysteriously began to flow again after a period of 70 years, forming a marsh once more. The record floods in 1925 had not caused a resumption of flow even although the Ngwezumba had flowed for a day or two, and for this reason its flow is believed to be connected with tectonic movements rather than climate. In 1965 the flow slackened and it was dry by October of that year; but the river filled again in 1966. In 1981 it suddenly stopped once more and remained dry for 30 years until water reached the Marsh again in 2010, and by 2022 it was still enjoying a wet period.

Oral evidence suggests the Mababe Flats may have held a lake in the eighteenth century. The Mababe River, which takes its origin from the Khwai River, fed a substantial swamp until about 1910, and it has held water on seven subsequent occasions.

8.8 An Ur of the Chaldees

In parts of South Africa the water table had fallen from 12–18 to 90–150 m by the 'fifties of the nineteenth century. But despite the pessimistic prognostications of many people since Livingstone's time that southern Africa was drying up, it was generally held the region was not undergoing progressive desiccation, but that rainfall simply fluctuated over time. But the analyses in support of the claim there is not progressive desiccation related to rainfall, whereas there has been a worldwide warming trend since about 1880 until the mid-1940s, when temperature in the southern hemisphere was almost one third of a degree higher than it had been in 1880. It then fell slightly but has been rising again since the mid-1970s. Higher temperatures mean greater and more rapid evaporation, thus increased dryness may not be detected from an examination of rainfall figures alone.

This rise in temperature has been verified on a local scale in Botswana. Analysis has shown the average maximum temperature at Mahalapye in the three months preceding the rainy season increased by a third of a degree between 1921 and 1946.

But at Ghanzi from 1951 to 1970 the rise was one whole degree due to habitat changes caused by overstocking.

Dr. Fritch, as early as 1868, considered the increasing dryness in South Africa was due not to overall changes in the climate but to the cutting down of trees. The principal cause of the drying-up of rivers and waterholes in the past two centuries is believed to be overstocking and overuse. In 1953 Professor Debenham warned of the need for careful management of cattle in such marginal country, referring to a cattle post near the Nata River "they had certainly made a mess of the country; the soil was trodden to bare sand within a mile of the cattle kraal and little verdure was left except a silvery shrub and a green broad-leaved weed, both inedible".

A fragile environment such as the Kalahari can only sustain sparse numbers of people and cattle. That it maintained such high numbers of game animals was due to their mobility and their varied means of exploiting the resources, from the giraffe's lofty browsing to the gemsbok's excavation of underground tubers.

Today in Botswana we are witnessing a stage towards the fate which befell Ur of the Chaldees, the Babylonian city which was abandoned after the Euphrates marshes dried up, leaving the ruins of a once great city in the middle of a desert; symptomatic of the desertification of the whole of the Middle East and North Africa in ancient times. A desertification that was caused by the destruction of trees and overgrazing of the land, causing springs and streams to dry up until nothing but desert land remained (Plate 8.3).

Plate 8.3 The Sahara of the south. Sand dunes at Bokspits

Chapter 9
They Had in Mind Their Children

Abstract Big game in Tswana culture. Hunting conventions, control by chiefs, chiefs' decrees. Protectorate government encouraged others from 1892. Control of firearms. Chiefs wish to retain control of game. Universal game law introduced in 1970. Early concerns at extent of slaughter. First statutory game law. In 1893 elephant, giraffe and eland protected outside of tribal territories. African views, tribal punishments. 1930s discussions with government. Chiefs' views.

Keywords Big game · Botswana · History · Attitudes · Legislation

9.1 Big Game in Tswana Culture

The wanton destruction of this once great game resource in Botswana was not due entirely to a lack of control, rather such control as there was proved ineffective. Game animals were an integral part of the culture of the Tswana. As a component of their initiation rites young Ngwato men had to hunt antelopes in the first year, all being obliged to prove he had killed at least one animal in a hunt before he returned from it. In the second year he had to kill either a rhinoceros, elephant, or buffalo. Even two of their calendar months are named after wild animals, *Phalane*, October, means literally "little *pallah*", denoting the month in which impalas calve. *Poane*, September, means "the month of young bulls", referring to the month when young male antelopes are expelled from the family herds.

Several tribes were distinguished by totem animals; that of the Kwena and Ngwaketse royal families was the crocodile, *kwena*. For the Ngwato and Tawana it was the duiker, *phuti*; of the Kgatla, *kgabo* the vervet monkey; of the Lete the buffalo, *nare*; of the Tlokwa the antbear, *thakadu*; of the Rolong the kudu, *tholo*; and the Kalanga respected the baboon, zebra, elephant, buffalo and eland. A totem animal could not be killed, or even touched, so protection was conferred on a range of animals.

Hunting was subject to a number of strict conventions. Mackenzie in 1881 referred to the "old tribal law" among the southern Tswana (now Bophuthatswana) whereby

C. Spinage, *Big Game of Botswana*, https://doi.org/10.1007/978-3-031-60033-3_9

the men of a tribe could go to a hunting station only with the chief's permission and then only for a short time, but the people now resented this control. Traditionally Batswana could hunt wherever they pleased within the Tribal Territory, even over the fields and grazing lands of others, without making any payment, or giving any tribute. But this right, in general, only applied to small game, or meat for the pot; even then qualifications existed concerning the age group of the hunters. Fur-bearing animals, however, were not usually hunted during the summer months when they were breeding.

In Tswana culture the ownership of game is not easy to define. On the one hand it was considered to be the property of the Chief, while on the other it was considered the Chief held it in trust for the Tribe and, as titular owner of the land, he was entitled to share in the proceeds of every hunting expedition. Thus a hunter could not usually dispose at will of game that he had killed and might be obliged to give the whole, or specified portions of it, to the Chief; or to the elders, or to certain members of his own family.

The Chief received all animals killed in a collective hunt he had organized (*letsholo*). Sometimes he divided the meat among those taking part, but usually suitable portions were dried and taken home for the people generally. The Chief kept the skins or sold them. Individual hunters were obliged to give the Chief the brisket (*sehuba*) of large game, the skins of lions and leopards, one tusk of an elephant (that of the side on which it fell), some feathers of an ostrich and the body of a bustard. This meant that every hunter returning from the hunt must carry a part of the produce to the chief.

The leader of a regiment received all animals or game birds killed while the regiment was in the field. If there was enough meat he shared it. A junior regiment was obliged to hand over such game to a more senior regiment. No one at a cattle post, or in the veld, could eat game birds, small antelopes, hares or tortoises, without first offering them to elders. The offering was passed on until it reached the oldest member, unless anyone gave the junior permission to eat it. Among the Ngwato, the head and neck of large game, and the marrow of certain species like kudu and hartebeest, had to be handed to the elders. These regimental and seniority obligations were still customary law in 1940, the latter still strictly enforced.

As a fundamental principle, the Tswana considered all fur-bearing animals to belong to them, as well as ostrich feathers; extending it about 1810 to ivory. This was irrespective of whom were the hunters, and Tswana rituals emphasized the trophies were tribal property. The Kgalagadi, San, Bayei, Tswapong and other formerly vassal peoples living in any district where there was a tribal overseer of one of the major tribes, were obliged to hand over all hunting spoils such as ivory, ostrich feathers and skins. A royal overseer kept such goods for himself, a servant overseer delivered them to the Chief. In 1875 Khama renounced such tribute; and in 1899 Bathoen allowed the Kgalagadi to sell skins and other products for themselves, but forced tribute persisted among the Ngwato and Tawana overseers with respect to the San until the late 1930s. As recently as 1935 the Kwena Chief insisted on his right to receive all skins of lion and leopard.

In the interior Livingstone noted the only game laws were that the man who first wounded an animal, although the wound was slight, was considered to be the killer of it. This, which was the same as Roman law, as the "first bullet" rule, became the rule of both indigenous and European hunters everywhere. The second person, Livingstone noted, was entitled to a hindquarter, and the third to a foreleg. The chiefs were generally entitled to a share as tribute; in some parts it was the breast, in others the whole of the ribs and one foreleg.

9.2 The Arms Race

Tswana chiefs laid claim to varying areas of land away from their villages and cattle posts which were solely hunting grounds. Along the Molopo river on either bank for a long way westwards of Mafeking, were the hunting grounds of the Rolong. North of these were those of the Ngwaketse, and stretching westwards into the Kalahari were those of the Kwena. Most extensive were those of the Ngwato reaching north-westwards to meet those of the Tawana.

The products of hunting provided the core of economic growth and political expansion of the main Tswana states in the third quarter of the nineteenth century. It is said their dependence on hunting had resulted from the disruption of settled agriculture caused by the period of instability in the first third of the century due to the *difaquane* or "forced migration", when plundering hordes of refugees, displaced by Shaka the Zulu king, moved westwards through southern Africa. The trade in game products, mainly ivory and ostrich feathers, became so important between about 1840 and 1870 due to demand in Europe and America before the ivory trade went into decline from over-exploitation, that the Tswana states competed and expanded to capture it. Around 1850 the centre for the trade was at Sechele's town Molepolole in the Kwena state, but it was gradually taken over by Sekgoma I at Shoshong in the Ngwato state, and the Ngwato hunted, or collected, skins from as far away as the Boteti. Baldwin in 1858 met a party which had "jackals, lynxes, wild cats, and skins of all kinds."

Guns, powder, and European clothing, were the most important trade goods bartered for ivory and feathers. Sought above all else were the guns and powder for defence and territorial expansion, especially in view of the memories of the *difaquane* and the fact the Matabele still posed a threat. Game therefore began to be slaughtered to satisfy an arms race.

Through an inability to maintain production Sechele lost out to Sekgoma, who in turn expanded his hunting activities, establishing greater control over the Bakgalagadi and the San employed as hunters. Khama III, who succeeded his father Sekgoma in 1875, consolidated this position; but by 1885, due to drought and over-hunting, there was insufficient game left in the Ngwato state also to support trade of any size. As the decline of the game could be seen in terms of direct economic loss, proposals for game laws by the British Protectorate administration were welcomed rather than resisted.

9.3 Control by the Chiefs

Outsiders were not permitted to hunt at will. The chiefs always objected to hunters entering their lands. Orpen reported opposition in 1852 from a Barolong chief and that the desire of the chiefs to keep hunting a monopoly for themselves pursued the expedition continually in the journey to Lehututu. In 1857 Sechele told Baldwin at Molepolole he would not have strangers coming into his country to hunt; but after a present it was allowed.

In 1875 Barber was prevented from hunting in the region of Makwa, about halfway between Khairi and Saasane, on the grounds it was a reserve for ostriches and Khama had given strict instructions that no hunters were allowed to camp there. When, in 1877, he asked Khama if he could collect young ostriches at Lake Ngami, (for stocking farms in South Africa), Khama "emphatically refused, saying that, as there were so few elephants left in his country, feathers were the only source of revenue left him; and if he allowed young birds to be caught, there would very soon be none left to grow up into big ones to grow feathers for his hunters to shoot." A hunter the previous year had taken out some young chicks, at which Khama was very much annoyed and warned the traders and hunters not to do so again. In 1886 Bethell was to write that he saw his first ostrich just beyond Sowa Pan.

In 1875 Khama sent a messenger to all the Ngwato farms to inform them that no hunters should remain at any watering-place for more than three days. This was due to the Afrikaners killing game indiscriminately for the sake of the skins, leaving the carcases to rot.

Khama informed Gillmore in 1876 that he had made up his mind not to allow traders or Afrikaners to hunt in his country, but for Gillmore, "a friend of Mr. Mackenzie's [the missionary at Shoshong], a soldier, and one who did not desire to exterminate the game, he would stretch a point." He granted Gillmore leave to hunt in his "own hunting-ground", Massara (Masarwa) veldt, a region north-east of the Motloutsie river and the Tuli river in present-day Zimbabwe. In fact a region claimed by Lobengula as his hunting ground.

The land between the Motloutse and the Shashe rivers was in effect a "no man's land" which had for long been a source of dispute between the Matabele and the Ngwato. Finaughty refers in 1869 to there having "always been a dispute" about the area. Anderson relates that Khama claimed the Shashe as his boundary line in 1877 but Lobengula claimed down to the Motloutsie river, and, indeed, all of Khama's country by previous conquest. In 1869 Mackenzie observed "The Tati district in point of fact has been and is reckoned the hunting-field of a head man called Sesinye, one of Macheng's people [chief of the Ngwato at that time]. As to whose country it is potentially, there can only be one opinion: that the Matabele could take it and occupy it when they liked, being stronger than the Bamangwato. In point of fact they have never done so; the only semblance of occupation was the herding of cattle by the people living on this line." Macheng claimed the Tati area on the grounds that his subjects had maintained cattle posts there for years, but in fact they had not gone beyond the Motloutsie and had been driven back from that river in 1863 by the Matabele.

Gillmore stressed that unless anyone had a proper introduction to Khama, then he would not get permission to hunt in his country, nor even to pass through it to hunt beyond. Khama had good reason to distrust the Afrikaners. Having granted three permission to hunt for ivory in one of his best game areas, after they had obtained a good quantity the elephants moved east into the tsetse fly country (in present-day Zimbabwe) where the Afrikaners could not follow for fear of losing their horses. So instead they slaughtered 700 buffaloes for their skins, leaving the meat to rot. When Khama heard of this he drove them out, having asked how dare they kill his cattle, "Were they not the food of his children when the heavens refused to give rain?"

Not only the Tswana wanted to exercise protection. Lobengula forbade any Afrikaner to enter or hunt in the Matabele territory under penalty of confiscation of all of his property. If he saw anyone eating ostrich eggs he would ask "how could you expect to get feathers if you ate the eggs?" He also denounced the killing of cow elephants and young ones.

In the 1880s Lewanika of the Barotse kept the Linyanti, at least on the northern bank, as an elephant preserve for himself; and in 1884 the Caprivi Strip north of Nambiya where elephants were plentiful was held strictly as a private preserve by king Lebuse.

9.4 Chiefs' Decrees

Bryden noted in 1890 that it was to Khama's credit he had set limits upon the "wanton destruction which for fifty years has been going forward". Afrikaner hunters were still not permitted and Englishmen also had to obtain permission. The tribal hunting parties which went out annually to collect skins were regulated by the chief, but owing to fear of attack by the Matabele, hunting expeditions to the Boteti, Mababe and Chobe regions had been few, and the game had increased. The exceptions were elephant and rhinoceroses which were "gone, or all but gone".

In fact control began much earlier and the first known chief's decree or *molao* controlling hunting dates to about 1815 among the Ngwaketse, when it was decreed ivory was the property of the chief. In 1856 Sechele forbade hunting on Sundays among the Kwena, probably in response to missionary prompting. Then in 1877 Khama forbade the capture of young ostriches, a tradition still guarded today by the Department of Wildlife with few exceptions. The following year he prohibited commercial hunting by Europeans, but allowed sport hunters upon personal application. This was followed by a number of other decrees, mostly encouraged by the Protectorate administration:

Year	Decree	Tribe
1892	Hunting of giraffe and other big game prohibited without permission of the Chief	Kwena
	Hunting of ostrich prohibited, but Chief gave permission to hunt cock ostriches	Ngwaketse
1893	Hunting of elephant, giraffe and eland prohibited without special permission of the Chief	Ngwaketse
1895	Hunting of giraffe, eland and other big game prohibited without permission of the Chief	Ngwato
1898	The use of deadfalls, staked pits and trapping on roads prohibited. Hoofed game to be caught only with jackal (iron) traps	Ngwaketse
1910	Hunting of elephant prohibited without permission of the Chief. Hunting of giraffe, buffalo, eland, rhinoceros and hippopotamus prohibited	Tawana
1913	Immigrants obliged to obey the Chief's laws concerning the destruction of game	Ngwato
	The killing of white storks and secretary birds prohibited	
	Hyrax and guinea-fowl are totally protected on Serowe Hill	
	"We must no longer kill big game animals"	Ngwaketse
1920	Elephants to be hunted only with the permission of the Chief, and one tusk to be given as tribute	Tawana
1926	Hunting of big game east of the railway line prohibited	Kgatla
	Setting traps in other peoples' fields prohibited	
1936	Sale of lion and leopard skins to traders prohibited	Kwena
1937	Hunting of giraffe and other Royal Game prohibited without permission of the Chief	Tawana

In 1913 it was reported the Tawana Chief gave very limited permission to people to hunt, except in respect of the smaller and more numerous species such as steenbok, duiker, impala and lechwe. By 1916 the hunting of large game was universally forbidden unless authorized by the respective chief.

With the disappearance of tribute, by the end of the 1930s people had more freedom connected with hunting than they had previously possessed under tribal law, but on the other hand, this freedom began to be restricted by the protection the chiefs extended to large game at the request of the Administration. Those seriously affected by such changes were the Kgalagadi and the San, and they frequently ignored the Chiefs' restrictions by not handing over the skins which the Tswana claimed as theirs. As for other tribes, hunting had become less important in the Tswana economy.

9.5 The Call for Greater Control

In South Africa the Sand River Convention of 1852, which gave the Transvaalers their independence, precluded certain Africans from obtaining arms and ammunition. This extended to Botswana and Sechele travelled to Cape Town to solicit the governor, Sir George Grey, to obtain exemption. The clause was abrogated in 1869 although not as a result of Sechele's visit. But at the beginning of the 1870s the people began acquiring guns and ammunition in immense quantities in payment for work in the recently discovered diamond mines in South Africa. Ironically, whereas this could have reduced the demand for game products for barter, it had exactly the opposite effect, and the first result was the destruction of the buffaloes throughout the valley of the Limpopo to the west of the Tuli river. But alarmed at the spread of fire-arms among tribes in South Africa which were becoming increasingly hostile, control was reinstated under the 1878 Peace Preservation Bill passed by the Cape parliament. This empowered the executive to issue proclamations ordering colonial Africans to hand in their arms.

Selous considered this restriction a mistake as far as the tribes in the interior of Botswana were concerned, who had used guns for elephant and ostrich hunting and to procure meat when the crops failed. The Ngwato he considered would suffer considerably as a result: "The trade in arms and ammunition ought certainly to be restricted within certain limits, but it is difficult to understand what good purpose its total suppression can serve in the far interior."

"Owing to the enforcement of this act, trade, the great medium of civilisation, is rapidly coming to a standstill in the interior".

Selous wrote that in 1881. But in 1897 in a letter to Sir Clement Hill who was preparing background material for the 1900 London Convention for the Preservation of Animals, Birds and Fish in Africa, he wrote: "The game in South Africa has been destroyed by three agencies: Firstly, by Europeans (principally Afrikaners of Dutch and French descent), who killed immense numbers of elephants and ostriches for their tusks and feathers; secondly, by native Africans possessed of fire-arms, who have been the principal factor in the destruction of game between the Limpopo and the Zambezi; and thirdly, by the rinderpest, which has been altogether beyond human control. With the last and most potent agency I have mentioned, it is impossible to cope; but the laws now in force in those parts of the interior of Africa, under the control of Europeans, ought to be sufficient to prevent the indiscriminate destruction of game by white men for the sake of their hides, and the comparatively small number of animals that are annually killed by prospectors and travellers for the sake of food-and the killing of which no laws will prevent in a wild country-will not do much harm. With the destruction of game by natives it is more difficult to deal, though it must not be forgotten that in many districts in the interior of Africa the destruction of game has been almost entirely due to natives possessed of fire-arms, whilst, at the present moment, it is hardly too much to say that 997 out of every 1000 elephants whose tusks come to the London market are killed by African natives ... The destruction of game by white men I think can be controlled and regulated, but it will be a much more difficult matter to prevent their extermination by black men, and the only way to do this is either by disarming them, or effectively preventing them from obtaining any kind of ammunition. It is absurd to make laws to prevent the killing of game by white men and at the same time to allow the native Africans to shoot them down at their pleasure. In my time, the white rhinoceros has been exterminated in northern and north-eastern Matabeleland entirely by native hunters-the small number of these animals killed in those countries by Europeans being a mere drop in the ocean; and at the present moment the game along the Upper Zambezi-where the game laws of the British South Africa Company are supposed to be in force-is being fast destroyed by the natives, who are well armed with guns and rifles, and who obtain their supplies of ammunition from the Portuguese settlements on the west coast".

The missionary Coillard agreed with Selous. He referred in 1897 to the fact that the "hunter's paradise" threatened soon to disappear. According to the elephant hunters near the Zambezi, the rinderpest had already made terrible havoc, "But it is the native hunters themselves, who, totally destitute of conscience in this respect, are hastening the extermination of certain species. It is high time that strict and intelligent laws should protect what survives".

Bryden considered that by 1890 Khama in time of war could muster 8000–10,000 men, "mostly armed with guns", guns which in time of peace were used for hunting. But in 1892 the Brussels Act, an international convention, prohibited the sale of fire-arms to indigenous Africans.

9.6 A Letter to Queen Victoria

In 1885 Khama sent a letter to Queen Victoria offering settlement of a part of his country in exchange for protection against the Afrikaners. Among other things he wrote: "My people enjoy three things in our country; they enjoy their cultivated lands, and their cattle stations, and their hunting grounds. We have lived [that is, survived] through these three things. Certainly the game will come to an end in the future, but at present it is in my country, and while it is still there I hold that it ought to be hunted by my people ... What I wish to explain is, that my people must not be prevented from hunting in all the country, except where the English shall have come to dwell. My people shall be stopped by cultivated lands and the cattle stations of the English inhabitants of the country".

When the British Government, at the request of the Tswana Chiefs, assumed jurisdiction over Bechuanaland in the year that Khama wrote his letter, the first Assistant Commissioner was instructed "not to interfere with the Native Administration; the Chiefs are understood not to be desirous of parting with their rights of sovereignty, nor are Her Majesty's Government by any means anxious to assume the responsibilities of it." It became recognized official practice to encourage chiefs to use their traditional powers to make laws, rather than to legislate by enactment, and they were on several occasions urged to make their own laws for the protection of big game in their areas.

The exception to the rule prevailing in many other African countries, was that in Botswana the laws introduced by the Protectorate Administration thus applied to foreigners only, but reinforced tribal law, the chiefs introducing game laws at the request of the Administration. It was, however, inevitable that such laws would eventually come to be adopted for the country as a whole as circumstances changed. Nevertheless, it was not until as recently as 1979 the statutory game laws became of universal application.

Concern for game received early attention perhaps as a result of complaints by Major Grey of the Bechuanaland Border Police, stationed at Macloutsie. In a letter of April 1894 to the Administrator Sir Sidney Shippard, he wrote that between March 1893 and March 1894 members of the Ngwato sold over 600 heads of large game; one store alone at Macloutsie buying 348 comprising 78 kudus, 51 gemsboks, 38 roan antelopes, 29 sable antelopes, 13 waterbucks, 27 impalas, 84 wildebeests, 24 hartebeests, and four buffaloes. All of which he believed came from within a 50 km radius. These were adult heads and the figures did not include about 100 immature specimens. His opinion was that the tribesmen were responsible for 90% of animals killed, and urged protection for sable and gemsbok, which he thought would soon become extinct.

The Assistant Commissioner, the Reverend J. S. Moffat, replied as follows:

Palapye
3rd July 1894

Sir,

"I have the honour to acknowledge receipt of your letter No. 5 a/c of the ninth of June enclosing one from Major Grey on the subject of the destruction of game in the Protectorate. The question is not by any means a new one. I have long ago discussed the matter with the Chief Khama. He is himself substantially in sympathy with the aims of Major Grey and his people have instructions accordingly.

It is only fair however that I should mention certain circumstances which have much discouraged him and his people in any serious attempt to preserve the game. Last year there was a great trek of Boers through the western part of the Bamangwato country, who were on their way to Damaraland via Lake Ngami. These people besides almost coming to blows with the herds at the Bamangwato cattle posts for the possession of the water-also hunted far and wide along their route and destroyed much game including many giraffes.

A man named Van Rooyen, who was at the time living in Tati, made excursions over the Shashe, into what is known as the disputed territory, and killed much game including giraffes. He was warned off by Khama, but not until the mischief had been done. Similar raids have been made over the Crocodile River by Boers from the Transvaal, who when Khama's people objected to their proceedings stated that they were there with the concurrence of the police detachment at Ngwapa. The police authorities have since denied this. I give these statements on Khama's authority as derived from his people and of course I do not vouch for their absolute correctness, but the effect is the same. Even with the Proclamation of September 19th last, there is much difficulty in enforcing the law. I think it is right to state all this to show or to account for the attitude of the native mind and for a certain amount of apathy on questions of this kind, but I repeat that Khama himself is in sympathy with the spirit of Major Grey's remarks: and is so far as in him lies carrying out a policy tending to a moderate preservation of the game."

In September the Assistant Commissioner at Gaborone, W. H. Surmon, then approached the Chiefs Sebele (Kwena), Bathoen (Ngwaketse) and Linchwe (Kgatla), on the subject. Chief Linchwe informed Surmon his people only killed game for food and he would be glad if no new law was made on the subject. In a letter to Chief Bathoen, Surmon asked that a close season be observed, and there be complete protection of elephant, eland, giraffe, gemsbok and sable; also young or half-grown animals and females. Bathoen replied the species no longer occurred in his area. Sebele answered by letter as follows:

"It has been my wish for some years that a close season should be observed re species of rare game … My people, the true Bakwena, do not agree with my ideas on the subject and have already said that my ideas are far too strict re game. But with your assistance I shall really do my best in this matter."

As a result of his approaches Surmon wrote to Sir Francis Newton, Colonial Secretary at Vryburg, in October 1894 as follows: "I think it would not be advisable to make any alterations in the game law at present and that it would be better to request the Chiefs to carry out His Honour's proposals amongst their own people as far as practicable in their own territories."

Hence, in a letter from the British Colonial Office of the 7th November 1895, to the Chiefs Khama, Sebele and Bathoen, which laid the foundations of the Protectorate, it was stated:

"Outside of the boundaries now laid down for the Chiefs, the British South Africa Company will administer; but the Chiefs will continue to have the hunting rights which they now enjoy, provided that they agree to observe a 'close season' and that they will nominate certain hunters for each year, to whom hunting licences will be issued by the proper authority."

The rights thus appeared to be conditional, but the Chiefs did not query the provisos in their reply, and they were never insisted upon. Neither did they query the ambiguity of the statement which could be interpreted as referring only to hunting outside of the tribal boundaries, but it has always seemingly been construed as referring to hunting within the tribal territories. As such it became the foundation for the tribal hunting legislation, which, instituted in 1967 after Independence, remained in force until 1979; when all hunting legislation was finally amalgamated.

9.7 The First Statutory Law

Automatically with the 1891 Proclamation the first written game law to be instituted in the Protectorate, applying to foreigners only, was an 1886 Act of the Colony of the Cape of Good Hope, titled For the Better Preservation of Game. This restricted the hunting of certain species of wild animals termed "game", and licences were required to hunt, capture or sell them. Included was the quagga (as well as the zebra which was sometimes called quagga) which had already been extinct for 28 years and was never known to have occurred in Botswana. It was not dropped from the Botswana schedule of protected species until 1940, by which time it had been extinct for 77 years.

At the same time The Ostrich Export Duty Act of 1884 came into force. This imposed a tax on the export of ostriches and ostrich eggs. After a 1904 Act no person, native or otherwise, could kill the hen bird of the ostrich or "remove, interfere with, disturb or be in possession of the feathers or eggs of such birds without the permission of the Resident Commissioner".

In 1893, elephant, giraffe, and eland, were specially protected except for the members of any tribe killing them within their own Tribal Territory, where the chief was expected to exercise control. But in 1904 provision was made to issue special permits *gratis* for these species.

The early pre-occupation of the British Government with game and game laws to the extent they were one of the considerations of the 1895 settlement, resulted from the appalling destruction which had taken place in South Africa in the preceding fifty years: a destruction which threatened the Bechuanaland Protectorate likewise. The introduction of legislation in 1891 may be seen as designed primarily to curb the massive commerce in animal products, and has remained a curb on commercial exploitation to the present day. It could be argued, in contemporary vein, that this early concern for game preservation was in order to restore the once important trade in game products, and so allow Britain to derive an economic benefit for herself from the Protectorate. Nothing in the documents of the time suggests this was the motive. Britain's actions were engendered solely by a genuine concern to halt the slaughter; a concern already Africa-wide at the beginning of the twentieth century. Events in Botswana were simply a part of a much wider initiative but yet maintained a unique independence of approach.

9.8 African Views

In the Protectorate years the Chiefs took their responsibilities seriously and at the Chief's Court, Kanye, in 1919 an offender was fined an ox for killing a hartebeest; and in 1937 six offenders were each fined two head of cattle for killing an eland. In the Chief's Tribunal at Molepolole in 1935, an offender was fined ten head of cattle for killing more eland than he was allowed to by the Chief. After Independence, in 1986 on two separate occasions accused persons were fined Pula 60 (about £20) for killing a hartebeest without a licence. In 1929, when game was much more plentiful, they could have been fined an ox with a value of about 73 shillings, for the same offence. Twelve times the value of the penalty awarded in 1986.

By the 1920s game affairs were discussed between Batswana representatives, protectorate officials, and representatives of the white community, in the Native and European Advisory Councils at Mafeking. The session held in November 1932 was chaired by the lively Rey. Discussion concerned restricting the killing of fur-bearing animals by season, age and sex, that is, not killing females and young. The proposals for control meeting with some opposition from the Batswana councillors: "The principal point at issue seems to have been misunderstood," stated Rey, "namely, the fur trade. We are trying to help you develop the fur trade, to improve your karosses and other articles you make from furs … The object of this Proclamation is this-to forbid traders to sell the skins of animals of this kind, or to buy them, and forbid such animals being killed at this time."

Rey continued with acerbity: "That is the whole object, and I cannot for one moment imagine that there could be any objection by anybody to such an obvious safeguard to your own interests. After all, if you sold all your cows you would not have much more cattle, would you? It is the same idea."

The councillors were not to be put down. Sehako Pilane of the Bakgatla spoke up: "We understand what you have stated about animals … I think Your Honour will recollect, however, that you gave us rams at the show last year. They have all been eaten up by jackals. While I agree with Your Honour that the jackal makes very beautiful karosses for great people, nevertheless, seeing that he has destroyed our rams, how can we spare the jackal?"

Rey: "My answer is this: If you cannot take better care of your rams, then we cannot give you any more."

Pilane: "I thank Your Honour for the reply. We have good dogs, Sir. I have trained these dogs to protect anything that Your Honour may give us, such as rams, and I have also trained them to find the spoor of the jackal principally, but what must we do if it happens to be a female?"

Rey: "You must teach your dogs to learn the Proclamation; they are very intelligent dogs; they can take care of the rams, so they ought to be able to learn the Proclamation".

Pilane: "The jackal is a very cute animal, Sir. It is always near people. We do not kill vermin at times when the fur is bad; according to our custom we hunt for furs in the winter, but we feel enmity towards the jackal, and hunt him now, because when the grass is green he gets very busy."

Another speaker interjected: "I wish to ask a question. The discussion arose about cattle, and I understand that means were to be taken to get rid of the ticks. This is a question I did not understand properly."

Rey: "Well, we are talking about the preservation of animals at the moment, and ticks are not fur-bearing animals. You may always kill ticks; the female of the tick and the young of the tick of any age … just for the moment we are talking about *fur-bearing* animals, in order to improve your trade."

Another took up the subject: "He referred to the possibility of a law which prevented the killing of game. I am discussing the subject because, in so far as we Bamangwato are concerned, the Mathogele and Matsapa regiments were born after the game had been protected. When we were at Shoshong the Chief Khama fined a man to pay for killing an eland. Another man was fined five head of cattle for killing an eland, while another was fined seven head of cattle. I am pointing out that the control of game has always been in existence to the owner of the reserve and among our people. After we had moved to Palapye, the Chief gave orders that trees should not be chopped down during his rule. He fined a man two head of cattle for cutting a tree down in the village; ever since I was young the preservation of game has been in existence. When I hunt, the Chief gives me instructions not to kill too many head, Sir. He used to search wagons for hidden meat. I wish to point out that game preservation is an old rule amongst the natives."

Rey replied he was not talking about game preservation generally, but simply about the killing of female animals in the breeding season, and of young: "What I want to know is quite simple: I assume the Chiefs agree with the idea, and I want to know whether they will support the Government and help carry out the law."

Tshekedi Khama, always a thorn in the side of the flesh of the Administration, was brief: "Seeing your Honour requires a reply from the Chiefs, I want to state that as far as the Bamangwato are concerned, we oppose this law and say there is no necessity for it."

The arguments continued at length. Tshekedi expressed it metaphorically: "we have already ploughed this land, and we do not find it necessary for Your Honour to re-plough it. We do not say that Your Honour cannot say to the Chief, if necessary: 'I notice that you are not looking after game.' It is Your Honour's right, but we oppose this law because it takes that which is our own, putting it in the control of the Government."

Rey's reply was followed by Chief Bathoen II of the Bangwaketse: "From our point of view, this point is an important one. We maintain that our fathers have always protected game up to the present". Rey wound up by stating the Council seemed to be a little divided on the position and that he would draw his own conclusions. No law was subsequently issued on this matter.

At the Session held in March 1938 under Resident Commissioner Arden Clarke, discussion centred upon a proposed limitation upon the export of biltong, imposing a tax upon any amount over nine kg. This was opposed by Tshekedi who claimed the people did not export biltong for commercial ends, but gave it as gifts to their friends. Arden Clarke answered: "What we are aiming at is to stop our game from getting wiped out as it has been in many parts of the Union. This section is aimed at white people who come in, particularly to the Tuli Block and the Kalahari District, and massacre the game there."

It was finally agreed that with the permission of the chief a local would be able to export more than nine kg of biltong duty free.

The question of the fee on ivory and rhinoceros horn was then raised. Chief Moremi of the Batawana complained the export fee was higher then the sale price, he went on: "There is apparently no extinction of the elephant ... The price is even less than 2s..6d [per 0.45 kg] if it is found that the ivory is not so good. Most of us as it is are depending on ivory and when people go to kill elephants they get permission from the Chiefs. They cannot therefore just kill for destruction".

Another added: "One and two men cannot go and hunt an elephant. A man has got to take four to five men, because it is difficult to kill elephants. A Chief might give permission to kill three but a man can only kill one sometimes, and the value he realizes is only sufficient to compensate those who accompany him, and he gets nothing owing to the smallness of the price offered".

Tshekedi: "we understand that in olden times when a man went out elephant hunting if he killed an elephant one tusk he offered to the Chief and the other one belonged to him. That indicated that elephants were under the control of the Chiefs. In my territory there are many elephants at present, but they are not hunted because of the law of the Chief. There are so many of them that they destroy crops but they have been protected. This protection was effected by the original owners of the country who had in mind their children, some of whom are present now".

The debate moved to the prohibition on sale of giraffe hides and tails in a 1925 law. The Resident Commissioner explained: "When we put this in we felt that all the evidence is that giraffe are gradually being wiped out and our idea was to have them entirely protected".

Tshekedi claimed he protected giraffe in his territory. He continued: "In my time when I first had suspicions that this kind of law would be introduced Chief Bathoen took a hunting expedition into my country. He went in search of two species of game which he could not find in his Reserve. They were giraffe and eland. He was accompanied by people. I think he and his companions killed about thirty head of game. Afterwards I observed that some District Commissioners discussed this with me viewing it is as if it were destruction and some other high officials in Mafeking spoke to me about it. I gathered they thought it was destruction. I then expected that I would get a direct accusation from the Government that I was not protecting the game and I would have answered that charge. I would have explained that Chief Bathoen was a Chief to me and as such his was a special privilege".

After an undertaking by Chief Moremi that he protected giraffe in his Reserve, the Resident Commissioner stated he would amend the law so that it did not apply to the Native Reserves.

9.9 The Chief's Goats

In October 1961 a discussion took place upon which depended the future of Botswana's big game. The lengthy debate concerning a proposed game policy boiled down to maintaining game inside reserves and eliminating it outside because of the threat of foot-and-mouth disease, particularly in European-held farming areas.

The Administrative Secretary: "it is suggested here that in the European farming blocks it will probably become necessary to encourage still further the shooting out of all game and the discouraging of private game reserves … In fact it may be that the Law might have to be amended to enable Government, in certain circumstances, to actually prohibit the establishment of private game reserves …

"In the Tribal territories, it is suggested that it might be practicable to have concentrated local campaigns of destruction in areas where it has been shown that the cattle population is frequently under risk of infection because of its proximity to areas of high game density".

Interventions of the African members showed a much greater concern for the conservation of game than they have been given credit for at that time. Thus a Mr. Ratidladi commented: "I agree that game should be protected. It is one of those beautiful things in nature which express the importance of the Creator. We Africans sometimes wish we could call back the Africa of our great-grandfathers, which literally used to flow with milk and honey. The Africa in which the able bodied young men used to chase the lion from its couch. But that Africa is gone. It was a bloody Africa, harassed by inter-tribal wars … The idea of using Africa as a zoo for the world where nature could still be found at its wildest and the raw African as the zoological specimen, has lost ground and has been substituted by the belief that Africa, like other countries has also the right to take a place in the general growth of the civilized world".

However: "game by and large are not respecters of fences and that during outbreaks of disease infected game might, and undoubtedly do, dissimulate infection over vast areas during their migration … therefore I can say nothing other than that the destruction or drastic reduction of the whole game population is necessary, and that extensive national parks should be established where game will be systematically protected.

"It has been suggested that the game population in the Tribal Areas should not be completely eradicated because it is a source of food. I do not attach too much importance to this suggestion, because it is surely going to be defeated by the exigencies of modern life".

Another spoke: "I also oppose the total destruction or drastic reduction of wild life. Animals wherever they are, are so intelligent that they will not associate themselves with the cattle posts where there is a high density of human population … I do not think if we can suggest to the Veterinary Department or to the Game Department that our wild life should be destroyed we would be being wise. We know that our young generation who are born today, in 10 years time will not know what is meant by an eland or hartebeest or kudu. It is wrong of course for us to advocate total destruction of those animals".

Chief Bathoen: "We will first examine, Sir, the truth about the statement as to whether game is found in abundance because of the low density of the population in Bechuanaland. That may have been so many years ago, when at one time elephant roamed even in the very place where we now sit which is still called "*Boswelatlou*"- the place where the elephant died. But game has now gone a far distance because of the increase in population and the destruction caused by human beings. Somebody said a human being is the most destructive creature that God ever created-I agree with that man entirely. The game that we used to find only a few years ago in Bechuanaland is not there owing to the destruction caused by people …

"I like game so much that I have put down my rifle for the last five years. I do not shoot anything, even birds. I like to see them flying over my head. God meant these creatures to be there, and we work against God's wishes and against nature. I allow people to hunt game in the Bangwaketse territory sparingly because game has moved away from where it was before. Game is not in abundance any more and I feel that whatever game is left should be preserved to some extent".

In Ngwaketse springboks became known as "the Chief's goats" because Bathoen did not permit them to be shot.

Mr. Mosielele continued the debate: "I am totally against the total destruction of game. We human beings do not regard ourselves as animals just because we have the power of destroying them. We honestly feel that because we have the power we should misuse this power by destroying other animals, with less intelligence and less education than us. I do not think that it is right. So, Your Honour, for that reason I feel that they should be reduced only when they are troublesome as some hon. Members have already stated".

The most significant comments were to come from Mr. Khama, significant because he was later to become Sir Seretse Khama, First President of the Republic of Botswana. His less benign attitude was to influence Government policy for the next 20 years, a policy not of 'benign' but of deliberate neglect. After dwelling on the subject of foot-and-mouth disease and the difficulty of marketing beef in some overseas countries, Khama continued: "I have heard it expressed that we should allow these, at one time, dreadful animals to roam around the whole of the Bechuanaland Protectorate, regardless of the harm they may do to our economy … It appears that some hon. Members think they have a choice and in fact they have, they must make up their minds whether they would rather breed wildebeest and various other objectionable animals like that or breed cattle. I feel that if we must agree that certain species of game do transmit or help to transmit foot and mouth, we should do everything in our power to reduce them and stop this disease. I cannot see any reason why we should suddenly feel sorry for the poor animals …

"We do not want to destroy all our wild life. We are hoping that we can preserve it. After all we cannot, no matter how the hon. Mr. England [one of the speakers] feels-he is for the destruction of everything that has four legs, except the cow-no matter how he feels, we have to consider the fact that there are various people in this country who are very much attached to game-who have a soul, who appreciate beauty and love to see springbok jumping all over the veld, even if they jump into a herd of prime cattle, infecting them with foot and mouth disease …

"We have to take into consideration the opinion elsewhere of those people who love wild game, who like to see it bouncing all over the place, no matter how much damage it does … We can move the menace from the more settled places, from the places where there is a concentration of cattle. We can gradually drive it back into the game reserves".

Placing the matter in abeyance, the Administration Secretary ended the debate with masterly understatement: "I do not think there has been any general support for the idea of general extermination and for that I think we all feel grateful".

Chapter 10
Vanished Are These Glories

Abstract Livingstone's observations of slaughter. Horror of pit traps. Use replaced by guns. Game already rapidly decreased. Nineteenth century observations. Slaughter of giraffe. Survival of giraffe. Want of animal life "inexpressibly sad". Predictions of extinction did not come to pass. Threat no of human population increase and change of land use.

Keywords Big game · Botswana · History · Destruction · Quotations

10.1 Before the White Man

"It is a frightful scene. The men, wild with excitement, spear the lovely animals with mad delight: others of the poor creatures, borne down by the weight of their dead and dying companions, every now and then make the whole mass heave in their smothering agonies." Livingstone was describing the traditional game trap or *hopo* as used in the drought in late 1840s on the Kolobeng river by the Bakwena: "The hopo consists of two hedges in the form of the letter V, which are very high and thick near the angle. Instead of the hedges being joined there, they are made to form a lane of about fifty yards in length, at the extremity of which a pit is formed, six or eight feet deep, and about twelve or fifteen in breadth and length. Trunks of trees are laid across the margins of the pit, and more especially over that nearest the lane where the animals are expected to leap in, and over that farthest from the lane where it is supposed they will attempt to escape after they are in. The trees form an overlapping border, and render escape almost impossible. The whole is carefully decked with short green rushes, making the pit like a concealed pitfall … the hedges are frequently about a mile long, and about as much apart at their extremities".

Between 60 and 70 head of large game were killed at the different *hopo* in a week.

Prior to the introduction of firearms this was the most favoured method of taking game in Botswana, but the use of firearms had fortunately already led to a decline in its use by Livingstone's time, for perhaps 100 head of game could be taken at once, directed into it by the funnel-shaped brushwood fence as used in modern game

C. Spinage, *Big Game of Botswana*, https://doi.org/10.1007/978-3-031-60033-3_10

capture. Set up near a watering point herds would be chased into this funnel at night by throwing firebrands of smouldering baobab wood at them. Once in the trap they think they see an opening at the end, an opening which confronts the pit. The result is a chaos of writhing, trapped bodies, those on top trampling those beneath but in their turn being speared.

The Kwena and Ngwaketsi used the *hopo* in daylight on the plains. Often using the same pit year after year it was usually sited in areas where game was abundant just the other side of a slight rise so that it was hidden from view. Herds of game were driven openly down the brushwood funnel at speed while spearmen were stationed behind the brushwood fence to deter breakouts. The Kgalagadi also used them: "whole herds of zebras and wildebeests are massacred at once … That part of the country abounded with the pitfalls made by these and others of the Bakalahari. Many of these had been dug expressly for giraffe".

There were other, individual pitfalls, sited on game paths or round pans, either staked at the bottom, circular or oblong-shaped, sometimes with a ridge across the middle over which the animal would fall, straddled across it. "The pitfalls which they make are about eight feet deep with a bank in the middle, so that whatever game falls into one of them falls across the bank and cannot touch the ground on either side" wrote Baldwin. He had much trouble with these pitfalls south of the Makgadikgadi Pans in the Orapa area in 1858: "the game is entirely exterminated: guns, pitfalls and poisoned arrows have done their work, and last year's drought and famine have left the natives with nothing to live on but the spoils of the chase."

Shortly afterwards a bull elephant was caught in a pitfall: "This country is done for the sportsman; there are too many Kaffirs, Bushmen, and Masaras, all hunters, and the pitfalls do very great execution; they are well supplied with guns, powder, and lead from Walvish Bay."

Another description of the *hopo* was given by Anderson, who wrote that it was in use "many years ago", probably meaning the 1860s. The pits were still there when he wrote in 1887: "In those extensive open plains, where tens of thousands of the antelope species roam, a favourable spot would be selected, and from eight to twelve large pits dug, ranged in a row fifteen feet apart, the earth taken out to a depth of five feet, and thrown up between them, forming a steep bank; at the bottom of this pit, it would be divided into smaller pits, two feet in depth, leaving a wall of earth between each; these would be square, and three in a row. The full size of the opening would be about thirty by twelve feet, placed longways. These would occupy a considerable space; at the extreme ends a thick bush hedge would prevent the game leaping over, and several hundred men placed in addition to prevent the animals going round. When all was prepared, men would drive the game by thousands towards the pits, and as they were pressed on by those behind they made for the pits to escape, where they would fall in, and having no foothold in the small square pits above named, had to remain. Hundreds passing over them, also got fixed, until the pits were full".

He noted as many as 1200 animals were caught at one time.

In 1884 Schulz and Hammar, travelling up the south bank of the Chobe River, noted: "A remarkable feature of the country lying between Lejuma [Leshomo] and here [Ngoma] is the quantity of game-pits we daily passed." They continued: "Twelve feet long, four feet broad, and perhaps twelve feet deep ... covered over with sticks and leaves hidden on top by a thin layer of sand. A long fence is also necessary to direct the course of driven game in the direction required ... Amusing tales are often related of how, when lions or other dangerous game have been entrapped in these pits, the natives, divided between fear and hunger (for lion's meat is also eaten by them), are much embarrassed to encompass the death of their quarry without injury to the much-prized skin ... We were told that a final resort to suffocation by smoke is the most reliable alternative to gain this object".

After their use had declined, much earlier in the south of the country than in the north, at Molepolole, when the people became short of meat, 30–40 men armed with guns went out with oxen and waggons for weeks at a time, hunting west in the Kalahari and returning when the waggons were filled with biltong.

From time to time the Ngwato conducted massed hunts or *letsholo*, an assembly under the leadership of the chief for hunting certain animals employed in rain-making; or for hunting elephants, ostriches, or lions. Mackenzie considered the *letsholo*, which surrounded ostriches or antelopes, was more dangerous than a war party, because the natives treated guns as they would spears and simply fired into a circle, more often than not hitting their opposite number on the far side. Bain had witnessed a *letsholo* in August 1826 near Molepolole, conducted by Ngwaketse warriors preparing to fight the Kololo in Kweneng: "the slaughter among the quaggas [zebra] and wildebeests became terrible; seventy of those large animals fell pierced with assagais in the course of a few minutes and three fine eland".

The last great game drive was conducted by Khama in 1906 for the High Commissioner, Lord Selborne, held at Serowe with 18,000 native beaters. The beaters were sent out far afield several days beforehand and then converged on a platform upon which the High Commissioner was seated. It was only partly successful due to lack of organization and the game breaking back, but we are not told how many head of game were shot.

10.2 The Coming of Firearms

In the 1840s the country north of the Orange River was "black with game". The various Tswana, Matabele, Basuto, Hottentots, Korannas and San, with their bows and poisoned arrows and their spears, and even the *hopo*, made no inroad into the incredible numbers of animals. Decline began with the advent of the European and his firearms. Even then it was the introduction of the percussion cap that greatly accelerated the rate of slaughter, followed rapidly by improvements in precision and rifling, and finally the breech-loading rifle.

"It is now-unless some effectual method of preservation can be quickly introduced-absolutely certain that the complete extermination of the great game of Africa, America, and Asia is merely a question of years. And in Southern Africa especially, this period of extinction is very close at hand." wrote Bryden in 1890.

Bryden dated the beginning of the decline to Harris's memorable expedition. But the Afrikaners were thrusting their way north into the Transvaal and it was economic growth which was the principal cause of the slaughter. Elephants were killed for ivory, rhinoceros for horns, and antelopes, giraffe and zebras for skins, rather than for meat; and the quest for ivory set in train the other once ivory became scarce. Traders who set up their centres in places such as Shoshong were not simply going to close down if they could help it when they could not get enough ivory, they encouraged the supply of other items in its place.

But once firearms were obtained in large numbers it was to the weapons of the obscurer huntsmen that the great bulk of the game fell, the local hunters and the Griquas.

"How are you to patrol or preserve those silent, illimitable deserts, hundreds of miles even from a native town, where the game still lingers, and the black hunter-often armed now-a-days with a good Martini-Henry breech-loader-is at work? At the same time one can hardly blame the black hunter for slaying the game of his country."

"Guns are now plentiful in every part of the interior, and native gunners-next to the Afrikaner skin hunters, the most reckless destroyers of animal life-are incessantly at work. Nearly the whole of British Bechuanaland and the Protectorate, as far as Khama's country, has now been denuded of heavy game."

"Not until we passed Khama's town of Palachwe [Palapye], and entered the waterless tracts of the North Kalahari, did we see any quantity of game … Tsesseby and pallah are becoming very scarce; of roan antelope we only found spoor once or twice. The elephant had disappeared … The rhinoceroses, black and white, and the buffalo, had disappeared" bemoaned Bryden.

10.3 Past Glories

Livingstone claimed the banks of the Boteti swarmed with game when he and Oswell first reached it in July 1849, but by 1853 Andersson reported that due to the persecution to which they were constantly exposed, and the introduction of firearms, the numbers of animals had rapidly decreased and those remaining were wild and wary. Even so, Oswell had described its banks as lined with pit-falls "from end to end": "They make the game very wild. One animal falls in and alarms the whole herd."

Somehow the game survived these deadly traps more effectively than it did firearms and already in 1862 Baines was to write: "but what a contrast to the reports of the first travellers who passed through teeming herds of wild animals, and who never thought of killing domestic cattle, great or small, after reaching Lake Ngami. The Botletle only a few years since swarmed with elephants, and the deserted and desolate plains we have passed over were alive with everything. Chapman says he could not feed his people unless he killed a camel (i.e. giraffe) every week, or at least some beast equivalent to it, and now so reckless has been the persecution, that within a day after a wagon comes to an outspan, hardly a solitary buck is to be seen within a dozen miles of it."

Oswell remarked: "Beyond the Zouga [Boteti] the wonderful abundance of animal life is not maintained. There is game but not in large herds. The happy hunting grounds in my time began at the Molopo and ended at the Zouga."

Robert Moffat wrote to his wife in June 1854 of the country west of Gaborone: "All this country, and far to the south, was not long since [1824] swarming with game-cameleopards, rhinoceroses (four species), buffaloes, tsessebes, elks [elands] and many kinds-seen daily and almost hourly by the passers-by, and one would have thought from the immense extent of comparatively uninhabited country through which we have passed, they would have continued to abound for generations to come. The musket, the rifle and the fleet steed have made them not only scarce, but almost to disappear altogether. Were I to judge from the country through which I have passed since I left the Kuruman, I should say the game has been completely swept away. The cause of this may be given in a few words. A party of hunters go out professedly to procure skins, fat and meat. They will return and, when asked about their success, will likely reply that they have not been very successful, or have had a poor hunt and not been able to bring any, or at least very little, dried meat home with them. I have heard such from small parties, who had shot sometimes a couple of dozen cameleopards, and from fifty to a hundred and fifty elands (elks), besides other game. When the question is put, what in the name of wonder they do with all the meat and skins, etc., enough frequently to load their waggons three times over, they will sometimes say one thing and then another-that the meat rotted from rain, etc., or that there were many poor people, or the game was poor and lean, and not worth skinning … how often we have expressed our surprise at the number of game shot down in their hunts. English hunters and amateur sportsmen accelerate the work of extermination. They often shoot down these beautiful animals, whether they need them or not, and often in places where there is not an inhabitant and are soon devoured by the eagles and beasts of prey. The Nimrods then get eloquent at the evening bivouac … as if they were beasts of prey instead of lovely antelopes."

Moffat was one of very few to abhor the slaughter. Two days later he continued his letter to his wife: "Today Bo-Same and a couple of other horsemen went out in quest of game. I felt a secret wish that they might not fall in with many, lest they should shoot more than we need or our full waggon can hold. It has always grieved me to see game shot and left to be devoured by beasts of prey; and such is the bewitching spirit of hunting that these lively animals that inhabit and adorn the otherwise uninhabitable wilderness-and it needs adorning!-are mercilessly shot down to add to the number of the hunter's trophies … The hunters … had killed three giraffes … some meat was procured from a female, which was the fattest. I could not hide my displeasure at this wanton waste, and referred to the general complaint along the road of want of game which want we were now increasing, becoming hunters for hyenas, lions and eagles. The thing appeared perfectly plain, and only required to be mentioned. I went and looked at the nearest giraffe; there it lay, its splendid tall and unique figure prostrate, its fine large black eyes, only half an hour ago so bright, now glazed. There it lay, till the eagles of heaven should descend on it, or the wild beasts of the desert, attracted by its smell, come roaring or howling to tear it and devour. One is ready sometimes to ask, 'Why such a necessity?'"

Of the region around Kolobeng where Livingstone reported so much game seeking water there in the drought, already in 1850 Leyland reported that since the introduction of firearms into the interior: "The game is more wild and scattered on account of being hunted so much."

Anderson wrote of the country on the road to Shoshong "the whole country swarmed with game, lions and wolves [hyaenas]". There were hartebeests, impalas and zebras. "The ostrich was then plentiful". He considered in Khama's country up to 1875 game "as well as lions, wolves [hyaenas], and other beasts of prey, were much more numerous than at the present time". North and west of here and along the Botete River and in the region of the Makgadikgadi Pans, elephants, rhinoceroses, buffaloes and giraffes, were plentiful "but of late years many hunters have been for months and scarcely met with any."

Yet in 1879 Gillmore commented on the abundant spoor of game in the eastern region of the country south of Gaborone, considering the unsettled state of the country prevented the people from hunting as much as they had done so before, "and there is a second reason for game having been so much more numerous than in former years, viz. the sale and importation of gunpowder having been prohibited throughout the Transvaal." In the region of Letlhakeng in Kweneng District he considered the country had not been disturbed by firearms.

But in 1887 Mackenzie wrote: "Thus the guns, which the natives were so eager to obtain, have enabled them to shoot down the game throughout the whole of the country; and when besides guns-as in the case of Khama's men at Shoshong-the people have horses, then the existence of all large game is a question of a few years".

The year before, in his address to the Royal Geographical Society, Farini had spoken of "the palmy days [fifty years before] when a successful season meant from one to two thousand pounds profit in skins and feathers."

10.4 Slaughtering the Giraffe

The pans at Boatlaname and Lephephe were favourite localities for the giraffe. Anderson refers to shooting two out of seven 13 km from Boatlaname and Mr. Hume on another occasion shot three from the saddle: "It is a pity such beautiful animals should be destroyed merely for food and skins."

In 1884 Walter Kerr, out hunting with Selous north of Boatlaname wrote: "Selous and Kirton … soon brought down the two giraffes-two large bulls … they had run so far from the track of the waggon that we could not utilize the meat, although it would have been most acceptable, because we had a large number of hungry dogs. But we cut out the tongues, and also had one of the hearts".

"The slaughter of giraffes found in this country must have been very great. Within two years over a thousand hides of these beautiful creatures have been shipped to the colony, for the purpose of being made up into cattle whips for ox-driving. It was affirmed that at that time they had sufficient cattle whips to last for ten years to come."

"A few years since they [giraffes] were to be found at no great distance from Khama's old capital, Shoshong; now they are first encountered in the bush and forest region beyond Kanne, or Klaballa [Tlhabala]" wrote Bryden.

In 1890 the same person had written: "in fifteen minutes we had thus bagged four giraffe-three cows and a bull; a sufficiency of sport to satisfy the most ambitious hunter".

The giraffe was sought mainly for its hide in Bryden's time, a hide at Palapye commanding a price of from two pounds ten shillings to four pounds ten shillings. Not many years previous the hides of rhinoceroses and hippopotamuses were used for making whips, but the rhinoceros was practically exterminated and the hippopotamus became scarcer year by year, so giraffe skins were used to fill the demand. Bryden noted "a few years back" there was a dearth of suitable hides and the price of whips rose considerably. This led to parties of Afrikaner and local hunters flocking to the Kalahari "and scores upon scores" of giraffe were slaughtered. The market was overstocked and the price of hides tumbled, preventing the fate of the rhinoceros falling to the giraffe. Nevertheless "It seems a pity that for the sake only of whips and sandals, and to furnish the hunter with meat and an exciting form of sport, this stately creature should be exterminated from South Central Africa, as it bids fair soon to be".

Khama's hunters made an annual excursion to the Kedia region, each party taking from 12 to 20 giraffes for their meat and skins, using the skins for sandals as well as whips.

"... in these days, when game is vanishing, almost as if by magic, from the face of the earth, unless its flesh is actually and imperatively needed, this, the most rare and characteristic of the African fauna, should, in my judgement, be suffered to go unmolested."

"Utter extinction is for the present, therefore, happily not immediately assured. But as Africa becomes opened up year by year-almost month by month-these beautiful and defenceless animals must surely pass away. At the end of another century, our successors will probably, as they gaze at pictures of the extinct camelopard, marvel that so extraordinary a creature could have lingered so late into the world's history".

Protected since 1906, today almost 9000 are estimated to have survived. Not many in a country of almost 600,000 km^2. But 20 years ago there was considerable pressure from the hunting lobby to open up the hunting of this species.

10.5 Inexpressibly Sad

Bryden was impressed with the Limpopo River, "No wonder Gordon Cumming and other travellers have written in such ecstatic terms of its beauties. Even at the present day the game has not quite departed from the neighbourhood of this river. Koodoos and pallahs in the bush not far away; leopards, bushbuck, and smaller antelopes in the coverts fringing the river, are yet found".

But in 1877, returning from Zimbabwe, Barber stated travelling along the Limpopo "not an antelope could we find. A great trek of Boers, with their families and stock, en route for Lake Ngami, had passed north just before us and cleaned out every head of game, and what was not shot had been frightened away."

This trek had assembled at the confluence of the Notwane with the Marico river in 1875 preparatory to emigrating to Damaraland in northern Namibia. Having started out it was forced to return to the Notwane after reaching Shoshong because of the lack of rain, waiting again until 1877. While waiting it greatly diminished the game on the Limpopo.

After crossing the Shashe in 1876, Holub noted the country seemed "full of game, but not to the same extent as former years. The animals of which I saw most were pallahs [impalas], zulu-hartebeets [tsessebes], harrisbocks [sable antelopes], and zebras." When he reached the Ramokgwebana: "the district about the Rhamakoban is noted amongst elephant-hunters for its great abundance of game; giraffes, zebras, grey pallahs [roan antelopes], harrisbocks, gnus, hyaenas, and bustards are frequently to be seen, lions, ostriches, and rhinoceroses being by no means rare."

In April 1884 Walter Kerr, having crossed the Ramokgwebana river, about ten kilometres into Zimbabwe observed: "While here I scoured the country for miles in search of game, but without any luck, although in bygone days there used to be abundance of sport found in these parts. That was before the war of extermination had been fairly started. Now, through the common property of gunpowder and the familiarity of arms to the natives, this region has been laid waste. You may travel not only for miles but for days in the veldt without seeing a living thing, save a few birds and, perhaps, a duiker. By no means a long time ago the shrill trumpeting of the elephant might be heard echoing among the kopjies which bank the river on the north; hundreds of giraffes browsed and found shelter in the mopani forests; in short, almost every species of wild animal to be found in South Africa was common among the now silent groves which fringe the Inkwezi.

"The native hunters, with their rude-looking arms, have been the exterminators. The finely-finished, specially-made rifles of the keen sportsman have done little harm among the big game, compared to the havoc made by these imperfect-looking weapons, with their clumsy stocks covered with hide, and altogether resembling a gas-pipe with a frozen clod of earth at one end."

"The utter want of animal life was inexpressibly sad."

Schulz and Hammar in 1884 similarly bemoaned the loss of game: "All the country from Pretoria to Mongwato [Shoshong] was more or less wooded … a country formerly inhabited by large quantities of game: but now along the route anything like large game has been shot out or driven away by the traders, whose waggons constantly traffic backwards and forwards along this road".

Travelling north-west from Shoshong, they recorded: "we faced the dismal journey through the sandy, bush-grown wastes, alternating with stretches of lower-lying turf flats, sparsely covered with mimosa and other thorn-clad trees. Formerly mighty troops of game inhabited these wilds; but the demand for ivory, and the unmitigated greed of many hunters, have effectually cleared this country of anything that had a tusk, plume or hide which could be sold".

When in September they reached the region of Ncamasere on the south bank of the Okavango, they were informed the game was gone and there was none within a day's march of the river. "Occasionally a giraffe appeared, but generally made off again, scared away, if not killed by them. Hippo there were in plenty in the river".

At the peak of exploitation one firm alone in Shoshong was exporting £50,000 worth of game products annually, worth more than some £1.5 million in today's terms. But by 1885 the trade had already declined due to drought and over-exploitation, the combined exports from the six stores at Shoshong amounting to only an estimated £15,000 a year.

In 1846 the trader David Hume was importing each year from Botswana into South Africa 900 kg of ivory, about 200 karosses and 90 kg of ostrich feathers. Ostrich feathers in the nineteenth century were worth £40–£50 per bird and Leask reported "a good many feathers bought" at Shoshong in September 1865. The ostrich-hunting season was limited by the fact that by about mid-October the feathers were worn out and no longer of value, and due to their sharp sight and fleetness of foot ostriches are not easy to hunt. But about 1870 an Irishman named Riley developed an effective method of hunting them on horseback in the Kalahari, although Khama refused him permission to use it in Bamangwato.

In 1906 the value of ostrich feathers per bird had fallen 20 times but was still equal in value to about 50 bat-eared fox skins, or two pounds ten shillings, and they were unmercifully hunted by the locals. One ostrich, as well as providing a favoured meat, would pay more than months and months of trapping the smaller fur-bearing animals. In 1912 the ostrich-feather boom reached a new height, but by then many of the feathers were produced from farmed birds in South Africa. Perhaps it was for this latter reason that the ostrich survived, and in 2017, 60,000 were estimated to exist.

10.6 Vanished Glories

"[V]anished are these glories", wrote Bryden in 1890 of Cumming's account of the fauna which abounded 45 years before near Lobatse, "gone now are all the magnificent forms; and there remains to this fair land of the Protectorate, until Khama's country is reached, only a few koodoos and pallahs here and there, small duykers and steinboks, and upon the mountains rheboks and klipspringers, to remind one of the most glorious and abundant fauna that the world has ever seen.

"Out in the Kalahari, however, in the western and north-western part of Sechele's [Kwena] country, are yet to be found eland, gemsbok, and hartebeest in some plenty, as well as occasional giraffe."

"I travelled through the whole of Bechuanaland, part of the Transvaal, much of the Kalahari and Ngamiland. Here I found the conditions of the great game even more deplorable in many ways than in Cape Colony ..."

"Nearly the whole of British Bechuanaland and the Protectorate, as far as Khama's country, has now been denuded of heavy game. Here and there in the former territory a few troops of hartebeest still eke out a shuddering existence … On the Molopo … there remained but a reed-buck or two …"

"To the westward, on entering the South Kalahari, I found gemsbok, hartebeest, koodoo, wildebeest, and a lion or two still maintaining a precarious sanctuary …"

"Not until we had passed Khama's town of Palachwe [Palapye], and entered the waterless tracts of the North Kalahari, do we see any quantity of game. In the North Kalahari, and along the southern bank of the Botletli river, Ngamiland, we found and procured specimens of giraffe, Burchell's zebra, eland, brindled gnu (blue wildebeest), lechwe water-buck, springbok and some of the smaller antelopes. Tsesseby and pallah are becoming very scarce; of roan antelope we found spoor once or twice. The elephant had disappeared … The rhinoceroses, black and white, and the buffalo, had disappeared. The lion is still plentiful along the Botletli river, but owing to the number of native guns, it is not often seen … The hippopotamus still survives, albeit in diminishing numbers, in Lake Ngami and the Botletli".

Of southern Africa three years later Bryden stated: "Nothing can well convey a sadder idea of the frightful waste of animal life that has taken place since the introduction of firearms into the interior than a perusal of the travels of Burchell … and Baldwin … and a comparison of the crowds of game they encountered with the present fauna. The game has not all gone, it is true, but the hunter nowadays, instead of having battalions of animal life blackening the plains around his waggons, has to trek very far and to work very hard to find sport. In another twenty years, unless preservation-a difficult matter in native territories-sets in strongly, African game in the countries between the Orange River and the Zambesi will be little else than a memory; and in one short century the matchless fauna of Southern Africa, which countless ages have looked upon, will be swept away."

"It seems probable that our descendants and successors will succeed to a world in which well-nigh all that is wild, natural, and beautiful has been crushed and improved out of existence, and from which almost every beautiful bird, every rare specimen of feral life has been eliminated."

"Fifty years ago, if any traveller or sportsman passing through the prairies of North America, then crowded with unnumbered millions of bison, or trekking over the plains of Southern Africa, at that time literally darkened by countless herds of many kinds of game, had been told that by the year 1890 nearly all of these magnificent creatures would have clean vanished from the scene, he would have laughed the idea to scorn."

Despite his pessimism, a United Nations report in 1962 considered Botswana was one of the most important areas for wildlife in Africa, holding the largest known concentrations of plains game. Somehow the big game of the Kalahari survived the overwhelming odds stacked against it and only one species became extinct, the white rhinoceros. In 1992 the black rhinoceros followed its congener into oblivion with the poaching of the last surviving bull. Can the rest of this once great fauna hold out?

10.7 Threats of a Different Kind

In 1895 it was estimated that the total indigenous population of Botswana was 60,000. In 1959 it was 300,000. Today the number stands at an estimated 2,352,000, up from 1,300,000 20 years before. In 1895 the number of cattle was estimated at over 1,000,000, in 1981 that number reached 3,000,000 and today it stands at about 1,9000,000 following drought losses. In 1950 the Colonial Development Corporation took out a lease on 39,000 km^2 of land in the north-east for cattle ranching on the grand scale. Designed to be the world's biggest single ranching enterprise, the Corporation calculated it would recover its investment in 25 years with the inclusion of an abattoir at Lobatse (there had been a cold-storage plant there since 1926), and over 800,000 ha of arable land in the Okavango Delta. Within four years the ranching scheme was declared financially unworkable. Only the abattoir was a success.

Yet in spite of the lessons of the past, and reservations expressed by successive studies, today organizations such as the European Community and the World Bank still try to develop grandiose agricultural schemes in these marginal areas which have already shown themselves incapable of supporting them.

The fences which have proved so disastrous to the Kalahari herds, the veterinary authorities argue, are not to stop the passage of foot-and-mouth disease, but to provide isolation cells for stock in times of disease outbreaks. In response to the criticism that the fences impede migration they claimed there was no proof the Kalahari herds do migrate. This is due to a conclusion reached from a one-year survey by international consultants in 1979, who doubtless based their statement that migrations did not occur on a comparison with the annual migration of the Serengeti wildebeest. It was a play on words, rather than a statement of fact.

A serious effect is that the fences are regularly patrolled by veterinary guards and any animals near to them are shot, so that even when the herds are not migrating there is a steady attrition of animal numbers. Accustomed to a continuous supply of free meat, even if there are no animals to be found near to the fences the guards search until they find an animal, albeit distant from the fence. Regardless of the veterinary restrictions which they were employed to uphold, the guards frequently transported the fresh meat across the fence line to be disposed of illegally. There is a story on good authority of a guard having shot a giraffe several km from the Kuke fence. Asked why he had shot the animal he replied with a big grin "Because it was walking towards the fence".

10.8 Words of Olive Schreiner

In 1938 the South African geologist A. L. du Toit warned the soils and vegetation of the Kalahari once destroyed were never replaceable. In 1960 the warning was repeated that although much of Botswana was capable of carrying a light stocking of cattle, the balance between maintenance and despoiling of the natural vegetation was extremely fine. But the warnings passed unheeded and that balance continues to be destroyed, destroyed by a stocking allegedly up to ten times above the recommended level. The result is the beauty of the Kalahari is fast fading forever.

The distinguished South African writer Olive Schreiner eloquently summed it up over a century ago in 1891: "there is a reckless and callous destruction now going on in our midst of that which can by no conceivable possibility be restored to humanity when once it is destroyed from the surface of the globe. Further generations will probably regard as intelligent and wise benefactors of their race the men who burnt the library in Alexandria and destroyed the Parthenon at Athens, when they are compared, in certain aspects, with the inhabitants of Southern Africa and the modern world.

"For the moment we are so entirely bent on advancing the claims of material civilization, which we are inclined to regard as the all in all of life, that more subtle, if equally practical and important, considerations are apt to be forgotten."

Despite its tragic past the game has survived, and Botswana today can offer unforgettable game viewing experiences for the visitor and is ranked among the top wildlife tourist destinations. In 2019 the national income from the tourist industry, which is wildlife based, outstripped that from beef exports by almost fifty times (Plate 10.1 and Fig. 10.1).

Plate 10.1 'The Lion Hunter'. Staffordshire figurines of Gordon Cumming

Fig. 10.1 Changing the face of Botswana. Upper: Hunting Advertisement 1892. Lower: Railway Advertisement 1898

Appendix A
Big Game Species of Botswana

Chacma baboon	*Papio ursinus*
Aardwolf	*Proteles cristatus*
Brown hyaena	*Hyaena brunnea*
Spotted hyaena	*Crocuta crocuta*
Cheetah	*Acinonyx jubatus jubatus*[a]
Leopard	*Panthera pardus pardus*
Lion	*Panthera leo*
Wild dog	*Lycaon pictus pictus*
Cape hyrax	*Procavia capensis*
Elephant	*Loxodonta africana africana*
White rhinoceros	*Ceratotherium simum simum*[b]
Black rhinoceros	*Diceros bicornis chobiensis*[b]
Burchell's zebra	*Equus burchelli burchelli*[c]
Chapman's zebra	*E. b. chapmanni*
Damara zebra	*E. b. antiquorum*
	E. b. antiquorum × *E. b. chapmani*
Bushpig	*Potamochoerus porcus*
Warthog	*Potamochoerus aethiopicus sundevalli*
Hippopotamus	*Hippopotamus amphibius capensis*
Giraffe	*Giraffa camelopardalis angolensis*
Klipspringer	*Oreotragus oreotragus transvaalensis*
	O. o. stevensoni
Common duiker	*Sylvicapra grimmia steinhardti*
	S. g. splendidula
	S. g. transvaalensis
Oribi	*Ourebia ourebi rutila*
Steenbok	*Raphicerus campestris steinhardti*

(continued)

C. Spinage, *Big Game of Botswana*, https://doi.org/10.1007/978-3-031-60033-3

(continued)

	R. c. steinhardti × *R. c. capricornis*
	R. c. steinhardti × *R. c. kelleni*
Sharpe's steenbok	*Raphicerus melanotis sharpei*
Kudu	*Tragelaphus strepsiceros strepsiceros*
Sitatunga	*Tragelaphus spekii selousi*
Chobe bushbuck	*Tragelaphus scriptus ornatus*
Limpopo bushbuck	*T. s. roualeyni*
Impala	*Aepyceros melampus melampus*
Reedbuck	*Redunca arundinum arundinum*
Mountain reedbuck	*Redunca fulvorufula fulvorufula*
Waterbuck	*Kobus ellipsiprymnus ellipsiprymnus*
Lechwe	*Kobus leche leche*
Puku	*Kobus vardoni vardoni*
Springbok	*Antidorcas marsupialis hofmeyri*
Gemsbok	*Oryx gazella*
Sable antelope	*Hippotragus niger niger*
Roan antelope	*Hippotragus equinus cottoni*
Eland	*Taurotragus oryx oryx*
	T. o. oryx × *T. o. livingstonei*
Buffalo	*Syncerus caffer caffer*
Blue wildebeest	*Connochaetes taurinus taurinus*
Cape hartebeest	*Alcelaphus buselaphus caama*
Tsessebe	*Damaliscus lunatus lunatus*
Vaal rhebok	*Pelea capreolus*[d]
Ostrich	*Struthio camelus*
Nile crocodile	*Crocodylus niloticus*

[a] The King cheetah, formerly known as *Acinonyx rex*, is simply a variant or dimorphic form
[b] Original populations extinct in the wild
[c] Extinct
[d] Occurrence never confirmed

Appendix B
Some Pioneers of Botswana

1807–8	EDWARDS, Rev. W. Missionary. Visited the Ngwaketse and traded ivory from them but left no record of his travels
1820	CAMPBELL, Rev. John. Missionary. Made a journey to Kaditshwene in 1820 passing through the south-eastern tip of Botswana. The first European to take note of the existence of a vast inland lake, which he heard of in 1813
1824	BAIN, Andrew Geddes. Trader, explorer and geologist. First crossed the Molopo River near Bray in 1824. Made a trip to Molepolole in 1826
	MOFFAT, Rev. Robert. Missionary. One of the first white men to enter Botswana visiting a village near Kanye in 1824
1830	HUME, David. Trader. Travelled to junction of Limpopo river with the Notwane in search of gold, followed the Mahalapye to its source and then crossed north to the Macloutsie river. In 1833 he visited the Ngwaketse, Kwena, Kgatla and Ngwato, probably reaching Shoshong
1835	SMITH, Dr. Andrew. Doctor in the Medical Department of the British Army employed by the South African Literary and Scientific Institution to explore the interior and find Lake Ngami. With an expedition of forty members, six of whom were Europeans, in 1835 he almost reached Mahalapye but turned back on account of drought
1845	CUMMING, Roualeyn Gordon. 1820–1860. Son of Sir William Gidron-Cumming, bart. Sportsman. Visited eastern Botswana in 1845 to 1849 and provided one of the earliest descriptions of the country. Hunted mainly in the region from Lobatse along the eastern side to about Serule. Exhibited his trophies at the Great Exhibition of 1851, afterwards touring the country lecturing and exhibiting, and then opened a museum the collection of which was purchased by Barnum after Gordon-Cumming's death and destroyed in a fire. He commissioned three large Staffordshire porcelain figurines: one of him in Scottish regalia holding a dead lion by the tail, one of him standing on a prostrate dead lion, and one of him with a dog; which he presumably sold at his exhibitions
	METHUEN, Henry, H. Hunter and naturalist. Hunted in eastern Botswana at same time as Cumming, entering just over two months later

C. Spinage, *Big Game of Botswana*, https://doi.org/10.1007/978-3-031-60033-3

OSWELL, William Cotton. Sportsman and explorer. Made his first hunting trip as far as the Kaa country near Shoshong in 1845. Visited Lake Ngami in 1849 with Livingstone

1846 LIVINGSTONE, David. Scottish missionary and explorer. First came to Botswana in 1846. Discovered Lake Ngami in 1849 with Oswell, Murray and Wilson

WILSON, J. H. Trader. At Kolobeng. Guided Livingstone's party to Lake Ngami. He visited Shoshong in 1848

1849 EDWARDS, Samuel Howard. Hunter and trader. Visited Kolobeng in March 1849 and Lake Ngami with Wilson in 1850, and the present Chobe National Park with Chapman in 1855. Visited Ghanzi and Chobe with Wilson in 1856–58. From 1888 he was official postmaster at Tati

1850 LEYLAND, J. Naturalist and trader. Reached Kolobeng in December and visited Lake Ngami in 1851 with S. H. Edwards and J. H. Wilson. The first ornithological collector in Ngamiland

MOYLE, Henry J. Hunter and trader. Visited Lake Ngami in 1850 and was the first European to reach Ghanzi, crossing the Kalahari from the east in 1852

1851 CHAPMAN, James. South African hunter, trader and explorer. Made his first hunting trip to the Molopo River and to Shoshong in 1851. In 1852 made an extensive safari to the north of the Makgadikgadi Pans with Jan Viljoen, Willem Prinsloo, Petrus Jacobs, Martinus Swartz and another South African hunter; all of whom had also reached Lake Ngami the year before. Visited Lake Ngami with Thompson and Campbell in late 1853 and made several subsequent visits in 1860–63 making a trip from the west to the Victoria Falls with Thomas Baines. Hunted in present-day Chobe National Park in 1853 and 1855

GALTON, Francis. Sportsman and explorer. Reached Rietfontein from west coast with Andersson in 1851

GREEN, Frederick Joseph. Hunter, trader, explorer. Visited Lake Ngami with Captain Edward Shelly and Gervase Bushe in 1851, and made several subsequent trips. With his younger brother Charles Alexander Green, George W. Bonfield and John Mortar, visited Lake Ngami from west coast in 1858 and then travelled east to Matabeleland, when he carved "Green's Expedition 1858.9" on the baobab at Gutsha Pan

ROBINSON, William. Former schoolmaster turned elephant hunter. Hunted at Lake Ngami

SHELLEY, Capt. Edward. Sportsman and explorer. Nephew of the poet. Became a Baronet. Visited Lake Ngami in 1851. Attempted to reach Damaraland to hunt elephants by crossing the Kalahari in 1852 accompanied by Richard Orpen, but forced to turn back at Lehututu for want of water

SIMPSON, J. Trader and hunter. Visited Linyanti

VAN ZYL, Hendrik. Disgruntled Afrikaner politician. Apparently reached Ghanzi in 1851 crossing the Makgadikgadi Pans, where he settled as the first white man there and became a well known elephant hunter. Carved his name on a baobab tree at Gutsha Pan in 1851

1852 McCABE, Joseph. Hunter, trader and explorer. First to cross the Kalahari desert from the south from Kang to west of Ghanzi in 1852. Made subsequent trips to the Chobe River and Lake Ngami

ORPEN, Richard. Cape settler. Irishman, brother of the leading South African politician Joseph Millerd Orpen. Accompanied Shelley on his attempt to cross the Kalahari from the south-east to the north-west, turning back at Lehututu

READER, H. Trader at Shoshong. Moved to Lake Ngami in 1852. In 1860 was on the Chobe River and also at Mababe

1853 ANDERSSON, Charles John. Swedish trader, hunter, explorer and naturalist. Accompanied by George Bonfield in 1853 the first to reach Lake Ngami from Walvis Bay on the west coast. Ascended the Thaoge River. Made a second journey to the Okavango River in 1859

1857 BALDWIN, William Charles. Hunter and trader. After an unsuccessful trip to the region of Francistown for ivory in 1857, in 1858 he struck north-west and followed the Boteti River cutting across from about Xhumaga to Lake Ngami. In 1859 he crossed the Ntwetwe salt pans to the Gweta area and hunted between there and the Thamalakane. He then travelled around the eastern side of Sowa Pan and along the border to Pandamatenga in 1860 on the way to the Victoria Falls

1859 HOLDEN, Dr. Scientific traveller, visited Lake Ngami. He died at Gobabis in 1861 of malaria

KENNY, John. Servant of J. Chapman on his journey to Lake Ngami in 1859

PALGRAVE, William Coates. Sportsman and hunter. Visited Lake Ngami in 1859

POLSON, Arthur and POLSON, George. Hunters and traders. Reached Lake Ngami from the west in 1859. George continued east to the Matetse River

1860 MACKENZIE, Rev. John. Missionary and later administrator. Began in Botswana in 1860

1861 HARTLEY, Henry. Farmer and elephant hunter. Hunted in Ngamiland

1862 BAINES, Thomas. Artist and cartographer. Visited Lake Ngami and Victoria Falls from the west coast with Chapman in 1862

1864 ANDERSON, Andrew A. Formerly a magistrate in Cape Colony, travelled in eastern Botswana from 1864 to 1877

FINAUGHTY, William. Hunter and later, trader. First hunted elephants in eastern Botswana with Edward Chapman in 1864. His preferred hunting ground was east of Shoshong to Francistown, his last hunt in 1877 being on the Shashe and in Zimbabwe. He was a resident trader at Shoshong from 1871 to 1875

1865 LEASK, Thomas. Trader. Visited Botswana almost as far as Shoshong in 1865 with G. A. Phillips. Became an elephant hunter making a number of subsequent trips

STROMBOOM, Jan Oscar. Norwegian hunter and trader. First visited Shoshong in 1865 or 1866. Visited Lake Ngami in 1867 and made several subsequent trips, establishing a store and living there in 1873. He made various other trips until 1890, dying at Mafeking in 1892

1867 WOOD, Swithen. Trader. Brother of the hunter George Wood. Accompanied Stromboom to Lake Ngami and set up trading store there, remaining for some 25 years

1869 MOHR, Edward. German interested in geographical exploration. Botanist and entomologist. Travelled through eastern Botswana in 1869

1871	HEPBURN, Rev. James Davidson. Missionary. Arrived at Shoshong with Mackenzie in 1871
	WESTBEECH, George. Trader, opened trading road north from Tati in 1871, established trading base and store at Pandamatenga in 1872 which became a well known rendezvous
1872	SELOUS, Frederick Courtney. Hunter, trader and naturalist. Travelled through eastern Botswana in 1872 and made several subsequent hunting trips in northern Botswana, first visiting Chobe in 1874
1873	OATES, Frank. Explorer and naturalist. Passed through eastern Botswana in 1873 on his way to the Victoria Falls but died of malaria on the return journey before reaching Tati
1874	HOLUB, Dr. Emil. Czech doctor of medicine and naturalist. Visited Shoshong in 1874, made second longer trip along eastern border in 1875
1875	BARBER, Frederick Hugh. South African farmer and mine owner, hunter. Visited eastern Botswana in 1875
	McLEOD, Capt. Norman. Sportsman. Passed through eastern Botswana in 1875 and visited Chobe
1876	GILLMORE, Parker. Naturalist and writer. Visited eastern Botswana in 1876 and subsequently
	SHELTON, Henry. Trader. Began trading at Lake Ngami in 1876
1877	McKIERNAN, Gerald. American trader and hunter. Reached Lake Ngami via Ghanzi from the west in 1877
1878	COILLARD, Francois. French missionary. Visited Shoshong in 1878, later lived at Leshomo
	PINTO, Major Serpa. Crossed from west coast, travelled south from Victoria Falls in 1878 with Coillard
1884	HAMMAR, August. South African farmer. With Dr. Aurel Schulz explored Chobe River in 1884
	SCHULZ, Dr. Aurel. South African medical doctor. Visited Chobe River and Okavango with Hammar
1885	HUNT, William, alias Gilarmi A. Farini. Showman. Claimed to have reached Lake Ngami from Khuis in 1885 but is believed to have only reached Lehututu. Nevertheless an early traveller through this region
1886	BETHELL, Alfred J. Adjutant in the Bechuanaland Border Police. Travelled in eastern Botswana in 1886
1889	NICOLLS, James A. South African sportsman. Undertook a hunting trip to Lake Ngami accompanying the trader J. Stromboom
1890	BRYDEN, H. Anderson. Traveller and naturalist. Travelled in eastern Botswana and to Ngamiland in 1890
1891	DECLE, Lionel. Frenchman, traveller. Undertook scientific expedition on behalf of the French Government to South and East Africa to study ethnology and anthropology. Passed through eastern Botswana in 1891
1895	GIBBONS, A. St. H. Hunter and explorer. Explored Upper Zambezi. Passed through eastern Botswana in 1895 and on his return in 1896 witnessed the effects of the rinderpest

1896 LUGARD, Frederick Dealtry. Administrator. Employed by British West Charterland Company to lead mineral prospecting expedition in Ngamiland. Travelled from Mafeking to Khwebe Hills in 1896 where he established his camp

PASSARGE, Dr. Siegfried. Geologist on Lugard's expedition. Explored northern Kalahari Desert in 1896

1899 REID, Percy C. Explorer. Explored the Linyanti region

1904 HODSON, Arnold W. Sub-inspector in the Bechuanaland Border Police, travelled widely in the country. His first journey was to Lehututu in 1904

1925 SCHWARZ, Prof. Ernest H. L. South African geologist, known for proposing the theory of "greening" the Kalahari by diverting water from the Okavango Delta. Visited northern Botswana in 1925

Selected References

Anderson, A. A. (1887). *Twenty-five years in a waggon. Sport and travel in South Africa.* Chapman & Hall Ltd.

Andersson, C. J. (856). *Lake Ngami, or explorations and discoveries during four years' wanderings in the wilds of South Western Africa.* Hurst & Blackett.

Bain, A. G. (1949). *Journals of Andrew Geddes Bain.* In M. H. Lister (Ed.). The Van Riebeeck Society.

Baines, T. (1864). *Explorations in South West Africa. Being an account of a journey in the years 1861 and 1862 from Walvisch Bay on the Western Coast to Lake Ngami and the Victoria Falls.* Longmans, Green, Longman, Roberts and Green.

Baldwin, W. C. (1863). *African hunting from the Natal to the Zambesi, including Lake Ngami, the Kalahari Desert etc. from 1852–1860.* Richard Bentley.

Bethell, A. J. (1887). *Notes on South African hunting and notes on a ride to the Victoria Falls of the Zambesi.* The Field.

Bryden, H. A. (1893). *Gun and camera in southern Africa. A year of wanderings in Bechuanaland, the Kalahari Desert, and the Lake River country, Ngamiland.* Edward Stanford.

Bryden, H. A. (1897). *Nature and sport in South Africa.* Chapman & Hall Ltd.

Campbell, A. C. (1990). History of elephants in Botswana. In P. Hancock (Ed.), *The future of Botswana's elephants* (pp. 5–15). Kalahari Conservation Society.

Campbell, J. (1815). *Travels in South Africa, undertaken at the request of the Missionary Society.* Black & Parry.

Campbell, J. (1822). *Travels in South Africa, undertaken at the request of the London Missionary Society; being a narrative of a second journey in the interior of that country* (2 vols.). Francis Westley.

Chapman, J. (1868). *Travels in the interior of South Africa, comprising fifteen years hunting and trading; with journeys across the continent from Natal to Walvisch Bay, and visits to Lake Ngami and the Victoria Falls* (2 vols.). Bell & Daldy.

Child, G. (1968). *Report to the Government of Botswana on an ecological survey of northern Botswana.* Food and Agricultural Organization of the United Nations.

Child, G. (1972). Observations on a wildebeest die-off in Botswana. *Arnoldia, 5*(31), 1–13.

Child, G., Smith, P., & von Richter, W. (1970). Tsetse control hunting as a measure of large mammal population trends in the Okavango Delta, Botswana. *Mammalia, 34*, 34–75.

Coillard, F. (1897). *On the threshold of Central Africa. A record of twenty years' pioneering among the Barotse of the Upper Zambesi.* Hodder & Stoughton.

Comrie-Greig, J. (1985). The law of the jungle: A sordid safari in Botswana. *African Wildlife, 38*(5), 174–185.

Cumming, R. G. (1850). A hunter's life in South Africa. In *Five years of a hunter's life in the far interior of South Africa* (2 vols.). John Murray.

C. Spinage, *Big Game of Botswana*, https://doi.org/10.1007/978-3-031-60033-3

Decle, L. (1898). *Three years in savage Africa*. Methuen & Co.
Farini, G. A. (1886). *Through the Kalahari desert. A narrative of a journey with gun, camera and note-book to Lake N'Gami and back*. Sampson Low, Marston, Searle & Rivington.
Finaughty, W. (1916). *The recollections of William Finaughty elephant hunter, 1864–1875*. J.P. Lippincott Co.
Galton, F. (1853). *The narrative of an explorer in tropical South Africa*. John Murray.
Gibbons, A. St. H. (1898). *Exploration and hunting in Central Africa 1895–96*. Methuen & Co.
Gillmore, P. (1878). *The great thirst land: A ride through Natal, Orange Free State, Transvaal and Kalahari Desert*. Cassell Petter & Galpin.
Gillmore, P. (1881). *On duty*. W. H. Allen & Co.
Gillmore, P. (1886). *The hunter's Arcadia*. Chapman & Hall.
Gillmore, P. (1888). *Days and nights by the desert*. Kegan Paul Trench & Co.
Gillmore, P. (1893). *Leaves from a sportsman's diary*. W. H. Allen & Co.
Green, F. (1856–58). Narrative of an expedition to the north-west of Lake Ngami, extending to the capital of Debabe's territory, via Souka River, hitherto an unexplored portion of Africa. *The Eastern Province Monthly Magazine*, I, 252–257, 316–323, 385–392, 533–543, 595–601, 661–669, II, 14–19.
Harris, W. C. (1838). *Narrative of an expedition into southern Africa, during the years 1836, and 1837*. American Mission Press.
Harris, W. C. (1840). *Portraits of the game and wild animals of southern Africa*. W. Pickering.
Hepburn, J. D. (1895). *Twenty years in Khama's country and pioneering among the Batuana of Lake Ngami*. Hodder & Stoughton.
Hodson, A. W. (1912). *Trekking the great thirst. Travel and sport in the Kalahari desert*. T. Fisher Unwin.
Holub, E. (1881). *Seven years in South Africa: Travels, researches and hunting adventures, between the diamond-fields and the Zambesi, 1872–1879* (2 vols.). Sampson Low, Marston, Searle & Rivington.
Jackman, B., & Godwin, P. (1986). Wildlife driven to the wall in the Kalahari. *Wildlife*, 72–79.
Kolben, P. (1731). *The present state of the Cape of Good Hope* (2 vols.). W. Innys.
Leyland, J. (1866). *Adventures in the far interior of South Africa. Including a journey to Lake Ngami and rambles in Honduras*. George Routledge & Sons.
Livingstone, D. (1857). *Missionary travels and researches in South Africa*. John Murray.
Mackenzie, J. (1868). *Ten years north of the Orange River. A story of every day life and work among the South African Tribes from 1859 to 1869*. Edmonston & Douglas.
Main, M. (1987). *Kalahari, life's variety in dune and delta*. Southern Book Publishers.
McKiernan, G. (1954). In P. Serton (Ed.), *The narrative and journal of Gerald McKiernan in South West Africa 1874–1879*. The Van Riebeeck Society.
Methuen, H. (1846). *Life in the wilderness; or wanderings in South Africa*. Richard Bentley.
Moffat, R. (1842). *Missionary labours and scenes in southern Africa*. John Snow.
Mohr, E. (1876). *To the Victoria Falls of the Zambesi*. Sampson Low, Marston, Searle & Rivington.
Neale, H. (1934). In G. C. Shortridge (Ed.), *The mammals of South West Africa* (2 vols.). William Heinemann Ltd.
Nicolls, J. A. (1890). Travel and sport along the Botletle River and around Lake Ngami. *The Field, The Country Gentleman's Newspaper*, Feb. 22nd. (1939), 289; Mar. 1st. (1940), 325; Mar. 8th. (1941), 363.
Nicolls, J. A., & Eglington, W. (1892). *The sportsman in South Africa*. The British Colonial Publications Co. and Simpkin, Marshall, Hamilton, Kent, & Co. Ltd.
Orpen, J. M. (1964). *Reminiscences of life in South Africa from 1846 to the present day, with historical researches*. C. Struik.
Oswell, W. C. (1893). In C. Phillips-Wolley (Ed.), *Big game shooting* (Vol. I). The Badminton Library, Longmans, Green & Co.
Passarge, S. (1904). *Die Kalahari* (2 vols.). Dieter Reimer.

Paterson, L. (n.d.). *Studies on the population ecology of the African buffalo (Syncerus caffer) in Botswana*. Government Printer.

Pinto, S. (1881). *How I crossed Africa: From the Atlantic to the Indian Ocean* (2 vols.). Sampson Low, Marston, Searle & Rivington.

Ross, K. (1987). *Okavango jewel of the Kalahari*. Southern Book Publishers.

Schreiner, S. C. C. (1925). *The migratory springboks of South Africa (the trekbokke). Also an essay on the ostrich and a letter descriptive of the Zambesi Falls*. T. Fisher Unwin Ltd.

Schulz, A., & Hammar, A. (1897). *The new Africa. A journey up the Chobe and down the Okavango rivers. A record of exploration and sport*. William Heinemann.

Schwarz, E. H. L. (1920). *The Kalahari or Thirstland redemption*. Maskew Miller.

Schwarz, E. H. L. (1928). *The Kalahari and its native races, being the account of a journey through Ngamiland and the Kalahari with a special study of the natives of that area*. Witherby.

Selous, F. C. (1889). *A hunter's wanderings in Africa, being a narrative of nine years spent amongst the game of the far interior of southern Africa*. Richard Bentley & Son.

Shaw, P. (1985). The desiccation of Lake Ngami: An historical perspective. *The Geographical Journal, 151*, 318–326.

Smithers, R. H. N. (1971). *The mammals of Botswana. Museum Memoir No. 4*. The Trustees of the National Museums of Rhodesia.

Spinage, C. A. (1990). Botswana's problem elephants. *Pachyderm, 13*, 14–19.

Spinage, C. A. (1991). *History and evolution of the fauna conservation laws of Botswana*. The Botswana Society.

Spinage, C. A. (1992a). Gleanings of game affairs in the Bechuanaland Protectorate. *Botswana Notes and Records, 24*, 19–32.

Spinage, C. A. (1992b). Further gleanings of game affairs in the Bechuanaland Protectorate. *Botswana Notes and Records, 24*, 33–38.

Spinage, C. A., & Matlhare, J. M. (1992). Is the Kalahari cornucopia fact or fiction? A predictive model. *Journal of Applied Ecology, 29*, 605–610.

Tabler, E. C. (Ed.). (1971). *Travels in the interior of South Africa, 1849–1863: Hunting and trading journeys from Natal to Walvis Bay & visits to Lake Ngami & Victoria Falls*. James Chapman, F.R.G.S. Edited from the original manuscripts by Edward C. Tabler, M.S. F.R.G.S. Part 2. A. A. Balkema.

Tyson, P. D. (1986). *Climate change and variability in southern Africa*. Oxford University Press.

Williamson, D. T., & Mbano, B. (1988). Wildebeest mortality during 1983 at Lake Xau, Botswana. *African Journal of Ecology, 26*, 341–344.

Williamson, D., Williamson, J., & Ngwamotsoko, K. T. (1988). Wildebeest migration in the Kalahari. *African Journal of Ecology, 26*, 269–280.